KU-763-928

Precious Metals Investing

FOR DUMMIES®

by Paul Mladjenovic

WILEY

Wiley Publishing, Inc.

Precious Metals Investing For Dummies®

Published by
Wiley Publishing, Inc.
111 River St.
Hoboken, NJ 07030-5774
www.wiley.com

Copyright © 2008 by Wiley Publishing, Inc., Indianapolis, Indiana

Published simultaneously in Canada

No part of this publication may be reproduced, stored in a retrieval system, or transmitted in any form or by any means, electronic, mechanical, photocopying, recording, scanning, or otherwise, except as permitted under Sections 107 or 108 of the 1976 United States Copyright Act, without either the prior written permission of the Publisher, or authorization through payment of the appropriate per-copy fee to the Copyright Clearance Center, 222 Rosewood Drive, Danvers, MA 01923, 978-750-8400, fax 978-646-8600. Requests to the Publisher for permission should be addressed to the Permissions Department, John Wiley & Sons, Inc., 111 River Street, Hoboken, NJ 07030, 201-748-6011, fax 201-748-6008, <u>or</u> online at http://www.wiley.com/go/permissions.

Trademarks: Wiley, the Wiley Publishing logo, For Dummies, the Dummies Man logo, A Reference for the Rest of Us!, The Dummies Way, Dummies Daily, The Fun and Easy Way, Dummies.com and related trade dress are trademarks or registered trademarks of John Wiley & Sons, Inc. and/or its affiliates in the United States and other countries, and may not be used without written permission. All other trademarks are the property of their respective owners. Wiley Publishing, Inc., is not associated with any product or vendor mentioned in this book.

LIMIT OF LIABILITY/DISCLAIMER OF WARRANTY: THE PUBLISHER AND THE AUTHOR MAKE NO REPRESENTATIONS OR WARRANTIES WITH RESPECT TO THE ACCURACY OR COMPLETENESS OF THE CONTENTS OF THIS WORK AND SPECIFICALLY DISCLAIM ALL WARRANTIES, INCLUDING WITHOUT LIMITATION WARRANTIES OF FITNESS FOR A PARTICULAR PURPOSE. NO WARRANTY MAY BE CREATED OR EXTENDED BY SALES OR PROMOTIONAL MATERIALS. THE ADVICE AND STRATEGIES CONTAINED HEREIN MAY NOT BE SUITABLE FOR EVERY SITUATION. THIS WORK IS SOLD WITH THE UNDERSTANDING THAT THE PUBLISHER IS NOT ENGAGED IN RENDERING LEGAL, ACCOUNTING, OR OTHER PROFESSIONAL SERVICES. IF PROFESSIONAL ASSISTANCE IS REQUIRED, THE SERVICES OF A COMPETENT PROFESSIONAL PERSON SHOULD BE SOUGHT. NEITHER THE PUBLISHER NOR THE AUTHOR SHALL BE LIABLE FOR DAMAGES ARISING HEREFROM. THE FACT THAT AN ORGANIZATION OR WEBSITE IS REFERRED TO IN THIS WORK AS A CITATION AND/OR A POTENTIAL SOURCE OF FURTHER INFORMATION DOES NOT MEAN THAT THE AUTHOR OR THE PUBLISHER ENDORSES THE INFORMATION THE ORGANIZATION OR WEBSITE MAY PROVIDE OR RECOMMENDATIONS IT MAY MAKE. FURTHER, READERS SHOULD BE AWARE THAT INTERNET WEBSITES LISTED IN THIS WORK MAY HAVE CHANGED OR DISAPPEARED BETWEEN WHEN THIS WORK WAS WRITTEN AND WHEN IT IS READ.

For general information on our other products and services, please contact our Customer Care Department within the U.S. at 877-762-2974, outside the U.S. at 317-572-3993, or fax 317-572-4002.

For technical support, please visit www.wiley.com/techsupport.

Wiley also publishes its books in a variety of electronic formats. Some content that appears in print may not be available in electronic books.

Library of Congress Control Number: 2007943297

ISBN: 978-0-470-13087-2

Manufactured in the United States of America

10 9 8 7 6 5 4 3 2

WILEY

About the Author

Paul Mladjenovic is a certified financial planner, author, consultant, and national seminar leader. He is also the editor of the "Prosperity Alert," a free financial newsletter found at www.SuperMoneyLinks.com. His businesses, PM Financial Services and Prosperity Network (at www.Mladjenovic.com) have helped people with financial and business concerns since 1981. Paul achieved his CFP designation in 1985.

Since 1983, Paul has taught thousands of budding investors nationwide through popular seminars and workshops such as "Ultra-Investing with Options," "The $50 Wealthbuilder," and "Rescue Your Retirement."

Paul has been quoted or referenced by many media outlets such as Bloomberg, CNBC, and many financial and business publications and Web sites. As an author, he has written the books *The Unofficial Guide to Picking Stocks* (Hungry Minds), *Zero-Cost Marketing* (Todd Publications), and more recently *Stock Investing for Dummies,* 2nd Edition (Wiley). In 2002, the first edition of *Stock Investing for Dummies* was ranked in the top 10 out of 300 books reviewed by Barron's.

In recent years, he accurately forecasted many economic events such as the bull market in precious metals and energy, the decline of the U.S. dollar, and the mortgage-credit crisis. At press time, he has been warning his students and clients about the coming energy crisis, rising inflation, and the long-term problems unfolding with America's retirement crisis.

Dedication

For the angels in my life: Fran, Joshua, Adam, and for my mother, Anna, an angel that left us to touch the face of God.

Author's Acknowledgments

First and foremost, I offer my appreciation and gratitude to the wonderful people at Wiley. It has been a pleasure to work with such a top-notch organization that works so hard to create products that offer readers tremendous value and information. I wish all of you continued success! There are some notables there whom I want to single out.

The first person to acknowledge is Jennifer Connolly, my Project Editor. Calling her magnificent is just not enough. Her professionalism, expert guidance, patience, and kind nature helped me get this book done during a personally difficult summer. She is a true publishing professional who has been extremely helpful, understanding, and patient. Those words are not enough to express my thanks for her fantastic guidance.

My Acquisitions Editor, Stacy Kennedy, has been fantastic from start to finish. I thank her for her efforts and the vision to see this project through from idea to reality. May the folks at Wiley always appreciate this pro!

I send my appreciation to Sheree Bykofsky and Janet Rosen for their professional assistance and personal support during the entire project. Through the years they have been superb, and I look forward to more of the same in the years to come.

To my wonderful wife, Fran, for her love, support, friendship, and devotion. I thank her for the free "tips" on precious metals and the numerous offers to do research at the jewelry outlets. Her tireless efforts will probably put me in Visa and MasterCard's hall of fame.

My thanks to my technical editor, Noel Jameson, another true professional and a great editor.

To all the great publishing, production, marketing, and distribution folks at Wiley, thank you for your dedication and wonderful efforts to bring *For Dummies* guides to our readers.

Lastly, I want to acknowledge you, the reader. Over the years, you have made the *For Dummies* books what they are today. Your devotion to these wonderful books created a foundation that played a big part in the creation of this book and will for many more yet to come. Thank you!

Publisher's Acknowledgments

We're proud of this book; please send us your comments through our Dummies online registration form located at www.dummies.com/register/.

Some of the people who helped bring this book to market include the following:

Acquisitions, Editorial, and Media Development

Project Editor: Jennifer Connolly

Acquisitions Editor: Stacy Kennedy

Copy Editor: Jennifer Connolly

Technical Editor: Noel Jameson

Editorial Manager: Jennifer Ehrlich

Editorial Supervisor: Carmen Krikorian

Editorial Assistants: Erin Calligan Mooney, Joe Niesen, Leann Harney, and David Lutton

Cover Photos: © Stockbyte

Cartoons: Rich Tennant (www.the5thwave.com)

Composition Services

Project Coordinator: Lynsey Stanford

Layout and Graphics: Reuben W. Davis, Alissa D. Ellet, Joyce Haughey, Christine Williams

Proofreaders: Debbye Butler, Cynthia Fields

Indexer: WordCo Indexing Services

Publishing and Editorial for Consumer Dummies

   Diane Graves Steele, Vice President and Publisher, Consumer Dummies

   Joyce Pepple, Acquisitions Director, Consumer Dummies

   Kristin A. Cocks, Product Development Director, Consumer Dummies

   Michael Spring, Vice President and Publisher, Travel

   Kelly Regan, Editorial Director, Travel

Publishing for Technology Dummies

   Andy Cummings, Vice President and Publisher, Dummies Technology/General User

Composition Services

   Gerry Fahey, Vice President of Production Services

   Debbie Stailey, Director of Composition Services

Contents at a Glance

Table of Contents

Introduction

*T*he time for precious metals has come. Emerging from the doldrums of a two-decade-long bear market, this millennium is witnessing a historic bull market for precious metals. Yet, the public hasn't caught on . . . yet. As you read this book, you can catch on as well and hopefully before the crowd does.

In recent years, I have told my clients and students that precious metals are (and will be) a necessary part of a healthy and growing portfolio. I don't tell people that precious metals are great because I wrote a book; I wrote a book because precious metals are great. *For Dummies* guides have become the quintessential nuts 'n' bolts introduction to a popular or necessary topic. Why not precious metals because I think they are necessary and their popularity is strong and growing? The time is now and the place to be is in precious metals (and some related places, too, such as base metals).

About This Book

Over the years, I have read and reviewed many investing books and lots of stuff on precious metals, but I didn't see much that could offer enough information and guidance for anyone interested in precious metals, especially if you are a novice in the topic. You may have some headlines on gold or even some commercial from some precious metals firm selling gold coins, but what is there to educate you as you dive into this fantastic topic?

Just about everything a beginner (or investor with rudimentary knowledge) needs to know about gold, silver, and other prominent metals is found right between the covers of the book that you are holding right now. As is the hallmark for *For Dummies* guides, the topic is laid out nicely so that you won't have to read from start to finish like so many other books. If the only thing you are interested in is how to buy gold or what you need to know about investing in silver, the information is there, easily found and ready to be read and comprehended in minutes.

Fortunately, (for me and for you), this *For Dummies* guide is not a clunky, laborious read. The writing is not stuffy and loaded with academic or industrial jargon. I get to write the book in my own voice; otherwise I'd fall asleep myself! It's a fun way to find out more about precious metals, and you can apply the info from this book immediately.

Conventions Used in This Book

I've used the following conventions to make your read through this book a bit easier:

- ✔ *Italics:* Although you probably know most of the basic investing jargon, I put words or phrases in italics when I define them for you.
- ✔ `Monofont`: Whenever you see a Web address, it will appear in monofont. This makes it easy to distinguish between the entire address and the rest of the text.
- ✔ When this book was printed, some Web addresses may have needed to break across two lines of text. If that happened, rest assured that I haven't put in any extra characters (such as hyphens) to indicate the break. So, when using one of these Web addresses, just type in exactly what you see in this book, pretending as though the line break doesn't exist.

What You're Not to Read

Okay, so I may not be the wordsmith my brain has me cracked up to be, but that doesn't mean you can go skipping stuff in this book . . . oh, wait . . . you can actually.

Perhaps you're in a hurry to get ahead of the crowd and start investing in some precious metals, so you need to cut to the chase and don't want to be bothered with the stuff that doesn't apply to you. Well, although I'd love to think that you're hanging on my every word, you can skip two things:

- ✔ **Any text appearing with the Technical Stuff icon:** This text, while meaningful and interesting, won't hurt you if you bypass it.
- ✔ **Any text included in a sidebar:** Whenever you see text in a gray-shaded box, those are sidebars. Of course, I love precious metals, so I've tried to include anything and everything about them in this book . . . well, at least as much as my editor allows me to. But some things just aren't necessary for your complete understanding of investing in precious metals. I've tucked those things away in sidebars, which, while interesting, again won't hurt you if you skip 'em.

Foolish Assumptions

Dear reader, I make a few assumptions about you. No . . . you're not a dummy but you would like more information on the topic of precious metals. You have some very basic knowledge of investing, and you understand that diversification means considering investments beyond merely stocks or bonds. You understand that inflation is a problem of modern life and that investments that excel in inflationary environments are a good consideration for any long-term investor.

How This Book Is Organized

Hopefully the book is laid out for you to easily find exactly what you want to find and with enough detail to give you some important insight on the topic — but not too much to overwhelm or bore you. Fortunately, every chapter refers you to other sources (such as books, Web sites, or other chapters) that can offer as much detail as you need or want.

Part 1: Breaking Down Precious Metals

You need to find the real deal on precious metals — why they're such a good thing and why the future looks so shiny for this stuff. Chapter 1 gets you started on what's all the hubbub. Sometimes the brightest futures really have their strongest foundations in history. Precious metals have been a useful part of economic history since civilization got . . . well . . . civilized.

Once you see the significance of precious metals, you can see why a commonsense portfolio needs to be diversified with (at least) a small portion devoted to precious metals. (Chapter 2 gives you more on this.)

The beauty of precious metals goes beyond just having a pretty face or a lustrous quality. It's also about "beautifying" your portfolio. You find out more about the beauty and benefits of metals in Chapter 3.

Nothing worthwhile in this life is without risk. Precious metals are no different. You can't have gold and silver and other attractive investments dazzle you, and then ignore the ugly parts. Risk is part of life and it is part of investing as well. It is also the 800-pound gorilla in the world of speculating and trading so a whole chapter is devoted to nothing but risk in the world of precious metals. In that case, make Chapter 4 required reading so that profits are easier to achieve (not to mention being able to sleep at night).

Part 11: Mining the Landscape of Metals

Yeah . . . you're right, "precious metals" can be a vague phrase so I get specific in this part. Each of the major metals deserves its own spotlight so I devote a chapter to each in this part.

You can't talk precious metals without the king — not Elvis — gold. The allure and prominence of gold throughout history has put it at or near the pinnacle of investment success and it has become an ageless symbol of wealth. Find out more about gold in Chapter 5.

I love silver even though it has been dubbed the poor man's gold. Silver's performance in the coming years will probably shock most people so look at it closely in Chapter 6.

Most books stop at gold and silver and tend to ignore the rest. Not I. There are lots of goodies to be discovered, such as platinum and palladium. This is also pricey stuff with a bright future. Find out more about the platinum group metals in Chapter 7.

I couldn't do a book on precious metals without the hot commodity of uranium. Energy is and will continue to be a major problem needing major solutions. Uranium will keep glowing, so find out why it's untouchable (no kidding!) in Chapter 8.

When was the last time you read something exciting about zinc or copper? Stop scratching your head and don't frown because base metals are a hot investment topic. Get the profitable perspective in Chapter 9.

Part III: Investing Vehicles

The variety of ways that you can invest, trade and speculate in precious metals are numerous and span from very safe and conservative to aggressive and speculative. Precious metals can be a vehicle to generate capital gains or income. They can be used as portfolio insurance or as risky venues that could be very profitable. This part covers the gamut.

Buying physical metals directly in bullion form is easier than you think, and I think that serious investors should have at least a small portion of their wealth in physical precious metals. Find out more in Chapter 10.

Another great way to get into metals like gold and silver is through numismatic (or collectibles) investment. It can be a little tricky so make sure you read Chapter 11 before you decide to proceed.

Probably the easier way to play the metals in a brokerage account is through mining stocks. They can be very profitable but there are some things to watch out for. Chapter 12 gives you the lowdown on mining stocks.

Maybe investing or speculating directly in mining stocks is not your bag. No problem — you can find great alternatives, such as mutual funds and ETFs, that are more conservative ways to add metals to your portfolio. Chapter 13 provides some great insights in this area.

For those who want the "high-flying" market where phrases like "get rich" and "crash and burn" are as common as "please" and "thank you" then you might consider precious metals futures (see Chapter 14).

Don't stop there! There are more opportunities in the world of precious metals by using options. There is an option strategy for just about anyone, and you can find the details in Chapter 15.

Part IV: Investment Strategies

Okay. You know that precious metals can be good and you now know what ways there are to participate. Now what?

If you are going to trade metals for fun and profit, start at Chapter 16.

If you are getting into stocks, futures, or options, you need a broker so check out Chapter 17.

For those who want the short-term moves in precious metals and in markets such as futures and options, find out more about technical analysis in Chapter 18.

You'd be surprised to find out that political intrigue has a big impact on precious metals (and your potential profits). The skullduggery can be found in Chapter 19.

After you start raking in the big bucks, then, of course, you have to grapple with taxes. Chapter 20 is the place to start.

Part V: The Part of Tens

The Part of Tens reads like top-ten lists for common topics on precious metals. Whether you need reminders on mining stocks (see Chapter 21), rules for metals investing (Chapter 22) and trading (Chapter 23), or you just want to check out some ways to limit risk (Chapter 24), take a stop at the Part of Tens.

Icons Used in This Book

I include some handy icons that you may notice in the margin of the book. They point you to certain types of information, so be sure you know which is which.

I include some text that tips you off into certain directions — this icon makes sure you notice. These are not tips of the "Psssst mac, have I got a tip for you" variety. They are more like "hokey smoke…such a great tip!".

Although I'd like for you to remember everything I say, I do have two kids and realize that's a losing battle. However, if you see this icon, be sure to ingrain this info on your brain.

Just like I want you to remember everything I say in this book, I'd love for you to *do* everything I say, but again, having two kids, I can calculate the rate of return on that. But to truly stay away from pitfalls that can cause you serious financial harm, you should heed any warnings you see associated with this icon.

Just like any expert, I do have nuggets of knowledge that only Alex Trebek, my Trivial Pursuit teammates, and my mother could love. But I guarantee that after discovering the value of precious metals could have in your portfolio, you could fall in love with some of this technical stuff, too. However, if you prefer, you can skip the info associated with this icon. This is *the only* icon that points you to info that you can skip if you prefer to.

Where to Go from Here

At this point . . . browse! Check out the detailed table of contents and go straight to those chapters that pique your interest. This is not a novel that you need to read from start to finish. It is like opening your fridge and pulling out what interests you. As you watch the precious metals markets become more popular (for many good reasons), come back and discover more as well as get pointed in the right direction for more and better ways to profit from precious metals.

Part I:
Breaking Down Precious Metals

In this part . . .

Why precious metals? For answers a lot better than "why not?", check out the chapters in this part. You can find a little history and a lot of reasons why precious metals belong to a diversified portfolio here. I hope you do it with minimum trouble, too, so you can also check out a chapter on controlling risk in this part.

Chapter 1

A Compelling History

*L*ong, long before government-issued currency, such as the dollar or the euro, existed, people must have had something else to help them trade with each other. How did people buy stuff they needed? Before there was any kind of currency, there was bartering unless you were a barbarian and preferred plundering. Civilized merchants and consumers traded goods and services, but trade did get cumbersome. For example, what if the merchant selling food didn't really want your 47 pounds of lint in exchange for a head of lettuce?

To make commerce a little easier, the buyers and sellers in the marketplace slowly decided on what could be used to facilitate trade. They decided that something had to be used as a currency, and that currency had to be portable and widely accepted as a unit of transaction, a store of value, and a medium of exchange. Whatever they chose as currency needed to be something that performed the role of . . . money! For thousands of years, precious metals — primarily gold and silver — filled the bill nicely. Very nicely.

Gold and silver came to be recognized as precious, valuable, and desirable in virtually every nation across the globe dating back to the dawn of civilized society. Now, you might ask "What the heck does history have to do in a *For Dummies* book?" Actually, history is very important because it will impact your portfolio in the coming years. History has shown us that there were (are) two major types of currency: precious metals and manmade (or government-issued) paper currencies. This chapter explains why the vast majority of paper currencies lost their value and are now gone but precious metals are still . . . uh . . . precious, making them worth a long look from investors.

In this book, I even take a step outside the word "precious" because I don't want anyone to call me "baseless." So I give you a look at the great opportunities that base metals (such as zinc and copper) offer investors and speculators (see Chapter 9).

Mining the History of Precious Metals

Everyone has used paper currency or credit measured in paper currency, such as dollars, but precious metals have withstood the test of time as a "store of value" and as a "medium of exchange" while most paper currencies in history went kaput — precious metals experts always use such technical terms as *kaput*.

Understanding why less is more

Paper currencies have a big problem: They're "manmade." Precious metals, such as gold and silver, on the other hand, aren't. Depending on who you are, you may consider gold and silver as created by God or Mother Nature's money, but in either case, gold and silver aren't — and can't be — created by mankind. Yes, you can find and extract or mine precious metals, but you can't create them out of thin air. On the other hand, over the centuries paper currencies (also called *fiat* currencies) were created by simply running a printing press — government-approved, of course. These days, the money creation authorities can do so even easier using a computer!

But being manmade gives room for abuse and misuse. Because the primary creators of fiat currencies were governments, those governments (through their power to enforce and mandate) made fiat currencies the money of (forced) choice. Because man can make money, man can then make a *lot* of money. However, you incur risk by creating a lot of money: if you create too much of it, it can slowly become less valuable, which is known as inflating the currency. Keep in mind that money is a reference to something of value that is used in exchange for something of value (such as goods and services). Currency is essentially a form of money that is generally accepted by a society as a convenient way to pay for those same goods and services.

Money retains its value by being limited or scarce. So, if you make lots and lots of money then each successive unit of that same currency becomes less and less valuable. This flaw in manmade currency explains why most currencies in history became worthless, and this danger still exists today. Yet, throughout time, gold and silver have retained their value. I guess you really can't fool Mother Nature!

Giving the gold standard a gold medal

Now before you think I totally love gold and silver while totally hating paper or fiat currency, guess again — it's the competition between the two that I'm not so crazy about. History tells us that the middle ground is actually a great place to be: Backing up manmade currency with gold — called the *gold standard* — works. Some forms of currency was backed up by silver by usually silver was used directly (in coinage, for example) since it was less valuable than gold. In modern times (the 1960s and later) silver generally disappeared from circulation as a form of money and was replaced by coins made of base metals.

Throughout history, the strongest and most stable fiat currencies were backed up by precious metals. Having a gold standard in place made the currency stable and its ability to purchase goods and services also tended to remain stable. Problems usually occurred when the currency was taken off the gold standard. Just take a look at American history for an ideal example: When America was on the gold standard from about 1800 to the late 1920s, general consumer prices were stable. However, as America gradually got off the gold standard and subsequently started to increase the number of dollars circulating in the economy, prices started to skyrocket. Alan Greenspan pointed out in 2002 that consumer prices doubled in the immediate years following the abandonment of the gold standard and had quintupled by mid-20th century.

A great primer on the history of money is available at the venerable Ludwig von Mises Institute (`www.Mises.org`). There are both a video and audio (MP3) to explain in laymen's terms the role of money and the Federal Reserve (America's "central bank"). This primer presents an excellent explanation of why gold and silver are critical to our economy's well-being.

The most common financial collapse occurs when too much money is created (inflation) thereby debasing it, resulting in a currency collapse or *devaluation*. The U.S. dollar is currently being created ("increasing the money supply") at a record pace and its value is shrinking at an alarming rate. History tells us loud and clear that diversifying even a small amount into precious metals is a prudent move.

Going for the gold

Nobody knows the exact details regarding the origin of gold usage, but the use of gold as money goes back to ancient times. Gold became an ideal form of money because of its durability and easiness to carry. It quickly became the most widely accepted currency among many different societies.

Gold became widely used as money in the American colonies with the Coinage Act of 1792. It played a major role in the U.S. economy up until 1933 when President Franklin D. Roosevelt prohibited the ownership of gold by private citizens. This prohibition only affected gold assets that could have been used as a competing currency, such as gold coins, bullion, and gold certificates. (For more information on gold coins and bullion, see Chapters 10 and 11.) Imagine that you do the smart thing in accumulating gold to preserve your wealth during the Great Depression and the government ends up confiscating it. Let's hope that it doesn't happen again.

Fast forward to our times and to what is unfolding in our economy and financial markets and you see that the conditions are ripe for gold to return as a necessary element in not only investors' portfolios but for consumers in general. As currencies lose valuable across the globe, more sturdy forms of money will emerge and nations will return to what has worked for centuries; precious metals.

Seeing the silver lining

Silver over the centuries had a unique dual role: monetary (used as money) and industrial. Going back to ancient times, silver was very commonly used as money, whether as minted coins or as a backing to paper money (such as silver certificates). Since it typically had a much lower monetary value than gold, it was actually more commonly used in commerce since it was great for smaller transactions. After all, wouldn't it be silly to buy a candy bar with gold anyway? Silver also proved to be a very versatile and useful metal for industry. Because of this, silver actually has some outstanding qualities for investment-minded folks. (See Chapter 6 for more details.)

Mentioning other metals

No book on precious metals would be complete without mentioning the other metals that have such great potential for investors. Here are some to consider:

- ✔ **Platinum:** A very pricy metal with attractive prospects for investors and speculators. (See Chapter 7 for more information.)
- ✔ **Palladium:** The "other white metal" offers some affordable investment opportunities as well (see Chapter 7).
- ✔ **Uranium:** Is a spectacular precious metal that is a great way to speculate in the world of energy as nations the world over build nuclear power plants (see Chapter 8 for full details).
- ✔ **Base metals:** They may be cheaper than precious metals but don't discount their powerful profit potential (Chapter 9 gives you the exciting details).

Taking a Look at Track Records

Before you check out each metal's track record, keep in mind that precious metals undergo major multiyear bull or bear markets reflecting the overwhelming economic and financial factors of that era (see the section, "Grappling with Bulls and Bears," later in this chapter for more about those markets in specific eras). Metals, both precious and base, are solid considerations for investors and speculators to get involved with but the bottom line is really the profit potential. The simplest way to judge the future potential of something is to check its past performance: the track record. Since 2000, metals and their related investments have had an enviable track record. I get into the specifics in the following sections.

Gold

Gold is the quintessential precious metal. "Good as gold" is more than a catch phrase; it 's the essence of gold's performance in recent decades. Some years from now they may change that phrase to "as sensational as gold."

That '70s metal

The 1970s was a historic time for our country's economy as well as for gold. For decades leading up to this decade, gold had a connection to the U.S. dollar. Gold began the decade at the government imposed fixed price of $35.08. However, the controls on the gold were gradually removed as the Federal government devalued the dollar in 1971. By February 1973, the government devalued the dollar again and raised gold's price to $42.22. During this year, the dollar ceased to be tied to the price of gold; it was now allowed to float and compete with international currencies in a free market. This unleashed the price of gold and its bull market was off and running! By June 1973, the market price of an ounce of gold reached $120 in London.

In 1974, the gold market really opened up. Americans were now permitted to own gold and many countries such as Japan lifted restrictions on gold buying and selling. In 1975, trading in gold futures (see Chapter 14 for more details on futures) began at New York's Commodity Exchange (COMEX) and the price of gold was left to find its free-market level. Market demand drove the price of gold per ounce to $180 in early 1975 before temporarily dropping to $100 in late 1976. Gold then steadily zig-zagged upward until it hit $240 in mid-1978 before . . . again . . . it fell temporarily to $190 in late 1978.

1979 was the big year for gold. By any measure, it had a great performance from the beginning of the decade to this point. Especially since it was a tough decade for the stock market and other investment vehicles. By early 1979, it was evident to all that commodities in general and precious metals & energy in particular. The cocktail parties were soon abuzz with people talking about international tensions, economic problems and . . . precious metals. I'm sure that if "The Graduate" came out in that year that some boorish party guest would have told Dustin Hoffman the opportunity of the day . . . "gold." The yellow metal zoomed to an all-time $420 by the fall of 1979 before taking a final breather to $380. Within weeks, gold spiked up to its famous-yet-brief high of $870 in mid-January 1980 before it came crashing down like a rocket ship that ran out of gas at the worst possible moment. It was time to switch your money from gold investments to parachute manufacturers.

For gold, 1980 started an extended bear market (a long period of dropping prices). However, the 1970s was one for the record books. Those who invested early on and persevered through the roller-coaster ups-and-downs of the gold market were handsomely rewarded with some spectacular profit opportunities: Gold's ride from $35 at the beginning of the decade to its peak of $870 at the tail end of the decade. That represents a percentage gain of a whopping 2,386%. To contrast, look at the performance of the Dow Jones Industrial Average (DJIA), a major barometer of stock market performance. The DJIA started the decade at 809 (January 2, 1970) and ended it at 839 (December 31, 1979). That 30-point gain represents a pretty measly gain of about 3.7% (for a whole freakin' decade!). To be meticulous about it, I realize that I'm off a few weeks; gold's all-time high occurred on January 21, 1980. That month, gold started at $559.50 before it spiked to its all-time high and then settling month-end at $653. Anyway, I think you get the picture. For the late seventies, it was indeed "as good as gold."

Gold stocks go berserk

As gold raced towards its all-time high during the late 1970s (hitting $870 in an intra-day high January 1980), gold mining stocks went ape. That's right; there was (and is) more than one way to make money with gold. During the late 1970s, stocks of companies that mined the yellow metal saw their share prices rise far greater than conventional stocks. Seeing share prices triple and quadruple in the gold mining sector was a common sight. There were two types of gold-mining stocks: the large companies ("the majors") and the smaller companies ("the juniors").

Large mining stocks easily beat the stock market averages. Homestake Mining, for example, was one of the majors. Its share price went from under $5 in 1978 to over $25 in 1980. Four hundred percent up in about two years: not bad! As a category, the large mining stocks did better in the last two years of the 1970s than the entire stock market did all decade long. Junior mining stocks did even better.

Silver

Silver isn't just a second banana serving as gold's sidekick. Its past, present and future look just as shiny.

Hi-ho silver!

Dubbed the poor man's gold, silver did extremely well in the late 1970s. While the world watched the higher-profile gold market, more astute investors and speculators noticed silver in gold's shadow. On a percentage basis, it did far better than gold. Silver started the decade at under $2 an ounce. However, the market for silver was opened up in similar vein to the gold market. Silver steadily rose to $10 by the beginning of 1979. Speculative demand by investors push the white metal's price past $20. The major influence on silver's meteoric rise came from a single private source: the billionaire Hunt brothers.

It is now a part of investment folklore but it was an intriguing true story. The Hunt brothers along with two wealthy Arab investors formed the company International Metals Investment Company, Ltd., for the purpose of cornering the silver market. Silver quickly soared to nearly $50 by January 1980. The market was to change rapidly as regulators moved in. Since the buying primarily took place at New York's Commodity Exchange (COMEX), exchange officials took steps to reverse the price rise. COMEX raised margin requirements (explained in Chapter 15) and temporarily allowed only sell orders on silver. These new rules created forced liquidations and caused the price of silver to plummet. By March 1980, silver fell to under $11, a drop of over 78% from its all-time high in less than two months.

It was indeed a wild ride for silver at the end of the 1970s. As the dust settled, nimble silver investors and speculators (learn more about the difference between these two in Chapter 3) made some spectacular profits in silver. The metal went from $1.29 in 1970 to its zenith of $49.45 in 1980. The percentage gain for the decade was an astounding 3,733%, certainly a silver lining for anyone's portfolio.

The 1970s might have been a great story and a distant memory that might cause us to daydream about what fortunes we coulda, shoulda, woulda made. The wealth-building power of precious metals is behind us, right? Well . . . not quite. This writer believes that the conditions are ripe in our time for a possible repeat performance. Why do you think I'm writing this book? Stay tuned . . .

Silver stocks go to the moon

In the late 1970s, there were silver mining companies that, of course, had a lot of silver. Millions of ounces of the stuff. How well do you think they (and their shareholders) fared? By 1979-1980, many of them experienced legendary profits. Take Lion Mines, for example. It was a junior mining company that you could have picked up for only 7 cents a share in 1976. By 1980, it was worth a staggering $380 a share. No, that's not a misprint. In other words, if you had bought $184 worth of that stock in early 1976, your shares would have been worth one million dollars only 41/2 years later.

Other metals

Platinum's general price action and market fundamentals were not that dissimilar to gold and silver. Just as the 1990s (for example) were relatively quiet for gold and silver, it was a similar experience for platinum. Its price moved in comatose fashion around the $400 level (give or take $25) from 1992 to the end of the decade. From 2000 onward, it was a different story. Its price soared upward and ended 2006 at $1,118 per ounce. Since it started the decade at $433, it rose a very nifty 158% during that time frame. A similar story line happened for its lesser cousin, palladium, during the early part of the decade. (Platinum and palladium are covered in greater detail in Chapter 7.)

Palladium traded for most of the 1990s in the $100-$200 range until 1996. Market demand (palladium is primarily an industrial metal) surged for it in the second half of the decade, driving its price skyward to just over $1,000 (Jan. 2000) before crashing down to the $300-$400 range in 2001. By the end of 2006, it hit a respectable $323.50. Even after you factor in the tremendous up-and-down of the 1999-2000 period, the price was still nearly a double after 10 years ended 2006. The market fundamentals for palladium (and the other metals) look very attractive going forward (I am writing this in early 2007).

Grappling with Bulls and Bears

The 1970s showed us a decade-long bull market while the 1980s and 1990s were an extended bear market (it was generally a period of falling prices for precious metals). The factors that made the 1970s very positive for precious metals (especially gold and silver) are back with a vengeance in this decade.

As the saying goes, "History may not repeat but it does rhyme." History is an important tool in deciphering what to do today. If the 1970s, for example, were a period of high inflation and economic difficulty, how do people react? How will they react today with similar conditions? In a nutshell, the past gives us real-time information to help us make appropriate investment decisions with "cause-and-effect" conditions. For me, it was a big reason for writing a book on precious metals since their bull market is the "effect" from the causes (inflation, economic difficulty, and so on). Kapish?

The precious metals 1980–1999 bear market

More times than not, you'll see a bear market after (and before) a bull market (duh!). For gold and silver, 1980–1999 was such a time. Although there were some interesting rallies (brief periods of price upswings), these two decades witnessed a long, painful zig zag down. 1980–1982 was indeed a brutal period for the metals. People walking the streets would have to watch out for traders and speculators hitting the pavement like water balloons at a shriners' convention.

There was a batch of reasons for gold's plummet. The early 1980s witnessed a dropping inflation rate which in turn meant decreasing consumer prices. Gold plummeted from its all-time high in early 1980 to $297 in 1982. However, gold staged a brief rally where the price rose to $500 in early 1983. During the late 1980s the catalyst for driving the price up during its two brief rallies was not inflation, economic problems, or international tensions. It was a market phenomenon as more buyers came in because of more discoveries of gold as new gold-rich mines were found that sparked renewed interest.

The second rally in the 1980s drove the price of gold again upward to briefly kiss the $500 level (for the last time in the century) and then it continued its downward, long-term trend. In the 1990s gold was able to touch the $400 level on several occasions, but it ended the century under $300. (The New York spot market closing price on December 3, 1999, was exactly $287.80.) The new century brought better price action for gold investors and speculators.

For silver, those last two decades were especially trying. Except for a few brief rallies to the $10, $8 and $7 levels, silver traded most of that time in a narrow range from $3.80 to $5.50. It ended the century at $5.33. What can investors learn from this?

The greatest profits come from understanding that market conditions can change dramatically over a period of years and decades that offer bullish (or bearish) profit opportunities. Sometimes a market gets beaten up so much that it gets to a point where it ultimately has nowhere to go but up. When your investment or asset hit rock bottom (what better phrase for metals?) then investigate and see if the data suggests a good entry point. The years 1999–2001 was just such a period as both precious and base metals hit rock bottom and began a long-term, powerful up move that rewarded many early investors.

The precious metals bull market of 2000–

History isn't always about what happened thousands (or dozens) of years ago. It offers valuable lessons right here . . . in your own decade! The years from 2000–2010 are an inflationary environment. Every major nation has been inflating its currency and in 2006–2007 the pace has accelerated. This is not good. At all! Unless, of course, you are into precious metals. At the beginning of 2000, you could have bought gold for 288.50 (New York spot market price) an ounce and silver for about $5.34 an ounce. By the last business day of 2006, gold hit $636while silver ended the year at $12.85. Percentage-wise, gold went up 121% and silver 141%. In contrast, the Dow started the millennium at $11,497 and ended 2006 at $12,463 for a modest gain of 8.4%.

Gold and silver stocks fared even better. Goldcorp, for example, is a major gold-mining firm that could have been bought for $2.88 a share at the beginning of 2000. By the end of 2006, your stock would have been worth $28.44 a share for a long-term gain of 887%. Silver Standard Resources, Inc. is a major silver-mining company that you could have gotten that first crisp trading day in January 2000 for a paltry $1.25 a share. You would have seen it climb into nose-bleed territory at $29.49 a share for a whopping gain of 2,359%. Many gold and silver mining companies had phenomenal results similar (or close) to these results.

Investors who had even a small allocation of their portfolio in precious metals or their related stocks could have boosted their total portfolio returns very nicely. In this chapter I use gold and silver as examples because they are the first things that people think of when they hear the phrase "precious metals." But the topic does encompass much more than the two famous metals. As you read earlier in this chapter, platinum and palladium offer great opportunities and their future looks pretty shiny too (pardon the pun).

Chapter 2

Diversifying with Metals

Diversification is always seen as a positive move for investors. Every investor (where possible) should have a variety of investment vehicles in his or her portfolio for obvious reasons. Diversification helps to minimize risk as well as increase the chances of seeing your overall portfolio grow, and the time has come for investors to diversify with precious metals because the economic and financial environment for precious metals is better than ever.

Working with Rising Inflation

The U.S. and other major countries have been increasing the money supply at a blistering pace. The increasing money supply means rising inflation, which is bad news for many conventional investments (such as bank certificates of deposit, bonds, and other fixed investments).

As the money supply grows, this monetary inflation then shows up as price inflation (higher prices for goods and services), which ties into hearing folks say "the U.S. dollar is losing value." Of course, the more dollars you create, the less valuable each dollar is worth. Inflation, then, is a serious problem that could wreck a traditional portfolio. Precious metals then become a solid addition to your arsenal of wealth-building vehicles.

As inflation and other ominous trends unfold, precious metals become not only a good investment, but also a *necessary* tool in your overall wealth-building and wealth-preservation picture. Gold and silver, for example, make for excellent long-term inflation hedges.

Understanding the Versatility of Metals

Precious metals have the unique quality (for investors, anyway) of coming in a variety of "formats" to fit a variety of investor needs. Think about that for a moment. If you bought a stock, it comes in one format . . . stock! Well, it could be a paper certificate or a digital blip that you see on your computer screen when you visit your brokerage firm's Web site but essentially, the only way you could invest in stock is . . . stock.

Precious metals, on the other hand, can be had as a "paper investment" (such as through a certificate of ownership or a futures contract) or shares in an exchange-traded fund (ETF; see Chapter 13 for more details) or in the actual physical metal itself. Unlike many other investments, the form the precious metal takes does indeed have a bearing on its level of risk and appropriateness (more on risk in Chapter 4).

The most versatile precious metals are gold and silver (and sometimes platinum). Gold and silver can be held as coins or jewelry but other metals are difficult to keep in physical form. Platinum, although available, is very expensive and not as widely available in physical form. Uranium? Don't even think about it!

Reaching Your Financial Goals

What do you want to do with your money? Money, precious metals, and all the other things that are part of your financial landscape are really tools for living — nothing more, nothing less. In the world of investing, trading, and speculating, precious metals become a means to an end. What is that end for you personally: making ends meet this week, paying off the mortgage next year, becoming financially independent within ten years, or another goal? Regardless of the goal you choose, precious metals provide you with an excellent vehicle to help you reach your future financial goals.

In 2001, a bull market began to unfold when the general investing public didn't even notice: Gold was under $300, and by the spring of 2007 it was nearly $700; silver was under $5 in 2001 and in the same time frame went to about $14. This bull market should continue, making metals a worthy investment.

You can find many investments and strategies suitable for long-term investing and many designed for short-term results. In the world of precious metals, the longer the term, the better your chances become for building wealth. The short term is more appropriate for traders and speculators, but keep in mind that the short term is fraught with volatility and risk. So short-term and long-term goals should address three things:

  ✔ Your risk tolerance (see Chapter 4)

  ✔ Your investing style (see the section, "Discovering Your Investing Style," later in this chapter)

  ✔ What your objective or goal is

The following sections give you an idea of how precious metals can help you achieve specific financial goals.

Seeking appreciation

Precious metals are primarily a vehicle for capital gain. In other words, investors seek to buy it at a low price and sell it at a higher price later. Precious metals do not usually generate income. Yes, there are mining stocks that issue dividends but these dividends are not substantial enough to merit consideration for investors seeking income. Precious metals have proven to be great vehicles that appreciate during long periods of economic, political, and financial distress. A good example is the late 1970s but a great example is . . . right now!

Early in this decade (about 2001), precious metals and precious metals–related investments (such as mining stocks) started an impressive bull market. The beginning was, however, a roller-coaster ride that zig-zagged upward. Every 12 to 18 months there was a pull-back or correction in their prices of 10% to 30%. For those not familiar with the volatility in precious metals prices, it was a scary moment, but for those with some patience and fortitude, the rewards were great.

Gold, silver, and other precious metals have shown excellent price growth and have been among the top performing assets so far in this period. They easily beat out financial assets such as stocks, bonds, and certificates of deposit. In the wake of the bursting housing bubble they even beat out real estate (cool!). For the rest of this decade and probably well into the next decade, the environment for rising precious metals prices is very bullish. Precious metals will shine (pardon the pun), but again, some patience and fortitude are necessary.

Looking for the home run

There are investors seeking appreciation (aren't we all?), but some of you may be looking for only truly fantastic gains. I know personally of one case in 1999 where $3,000 was turned in $80,000 in only four weeks in a commodities bro-kerage account. The vehicle he used was options on gold futures. During that brief time frame gold rallied from $250 to over $300. Of course, another four weeks later the amount shrunk to $22,000 when gold went back down. Remember: Home runs can also easily turn into strikeouts.

For those looking for home runs, precious (and base) metals offer realistic opportunities for great gains during the coming years. Base metals have been very profitable in recent years, and they are covered in Chapter 9.

Precious metals are part of the natural resource sector, which includes not only metals but also commodities (grains, meats, and so on) and energy (uranium, oil, natural gas, and so on). Because metals are a finite, nonrenewable commodity whose supply is strictly limited to what we can find both above and below ground, it offers the chance for great gains since demand and usage for them keeps going up. The population keeps growing right along with the money supply. As consumer and industrial demand keeps going up and as we print up more and more dollars (along with euros, yen, and other currencies) then what do you think will happen to the price of finite resources? Can you say "powerful, long-term, profitable bull market" three times fast?

Preserving your capital

Keep in mind that precious metals are assets with intrinsic value; value that's been there for literally thousands of years through all sorts of economic conditions. In recent decades precious metals performed very well but in part it was due to what was going on around the economic and financial markets. Inflation is a dreadful problem. Inflation became more of an issue in this decade, and the prospects look bleak for the next decade. Why? Our government (and governments of the other major nations) is increasing the money supply at record rates. Coupled with persistent problems in areas such as, the war on terror, Social Security, healthcare, budget deficits — just to name a few — ensure (unfortunately) an environment conducive to precious metals. Precious metals have a long record of maintaining and rising value so they can be an attractive part of even conservative portfolios.

Using precious metals to generate income

Although precious metals are best suited for gains, there are precious metals vehicles and strategies that can generate income. This is important to keep in mind because no one says that once you are into a strategy to generate gains that you must stay that way. Au contraire! When your situation changes and you want to start reaping some of the fruits of your labor to augment your income needs then you can do something about it with these income choices:

- ✓ **Dividend-paying mining stocks:** If you have grown your money with small mining stocks and you are changing your strategy from capital gain to income, then you can sell your no-dividend stocks and use the proceeds to buy larger, established companies that offer a dividend. Some offer quarterly dividends and there are companies that offer monthly dividends.

✔ **Covered call writing:** This strategy could easily generate income of 5% to 15% or more from your existing stock portfolio with no added risk. Covered call writing could be done with the stocks of large mining companies no longer looking for gain.

✔ **Bonds issued by mining companies:** I include this to be complete but they are not that different from bonds issued by other companies. Don't consider it to be a "precious metals" investment; it's just a bond so don't get too excited.

✔ **Marketsafe certificates of deposit:** Issued by Everbank (`www.everbank.com`) this has the features of a regular CD (interest) and it's tied to precious metals such as gold and silver. If gold goes up, your CD (protected by FDIC) can increase in value with no added risk. Visit the Everbank Web site for more details.

Discovering Your Investing Style

When you think of style, perhaps you think of famous labels and famous people, but you probably don't think about those folks with their colored coats waving and yelling at each other on the stock trading floor. Regardless of what you may think, they've got style, and so do you — an investing style. In the following sections, you can nail down your style and match it up with the best vehicles for precious metals investing.

Distinguishing between styles

In my investing seminars I talk to my students about the differences in the concepts of saving, investing, trading, and speculating because it matters so much and because I find it unsettling that even financial advisors don't know the difference. These are important to distinguish because people should know otherwise the financial pitfalls will be very great. The differences aren't just in where your money is but also why and in what manner. Right now as you read this, millions of people are living with no savings and lots of debt, which tells me that they are speculating with their budgets; retirees are day-trading their portfolios; and financial advisors are telling people to move their money from savings accounts to stocks without looking at the appropriateness of what they're doing. Make sure you understand the following terms — knowing the difference is crucial to you in the world of precious metals:

✔ **Saving:** The classical definition of saving is "income that has not been spent" but the modern-day definition is money set aside in a savings account (regardless of the interest rate) for a "rainy day" or emergency. Everyone should have a savings account with money that is safe and accessible just in case you encounter an unexpected interruption in your cash flow. In fact, you should have at least three months' worth of gross living expenses sitting blandly in a savings account or money market fund. Do you know that 2008 is the year that the first wave of baby boomers (there are a total of 78 million baby boomers) start to retire? These retirees have more debt than any prior group of retirees and their savings rate is around . . . zero! Although precious metals in the right venue is appropriate for most people, including savers, you need to have cash savings in addition to your precious metals investments. A good example of an appropriate venue in precious metals is buying physical gold and/or silver bullion coins as a long-term holding (see Chapter 10).

✔ **Investing:** This term refers to the act of buying an asset that fits an investor's profile and goals that is meant to be held long-term (in years). The asset will always run into ups and downs, but as long as the asset you're holding is trending upward (a bull market), you'll be okay. Investing in precious metals may not be for everyone but it is an appropriate consideration for many investment portfolios. The common stock of large or mid-size mining companies is a good example of an appropriate vehicle for investors. (See Chapter 12 for more details.)

✔ **Trading:** Trading is truly short-term in nature and is meant for those with steady nerves and a quick trigger finger. There are many "trading systems" out and this activity requires extensive knowledge of market behavior along with discipline and a definitive plan. The money employed should be considered risk capital and not money intended for an emergency fund, rent, or retirement. The venue could be mining stocks but more likely it would be futures and/or options since they are faster-moving markets. (See Chapter 14.)

✔ **Speculating:** This can be likened to "financial gambling." Speculating means that you're making an educated guess about the direction of a particular asset's price move. You are looking for big price moves to generate a large profit as quickly as possible, but you also understand that it can be very risky and volatile. Your appetite for greater potential profit coupled with increased risk is similar to the trader but your time frame is different. Speculating can be either short term or long term. Your venue of choice could be stocks, but more likely, the stocks would be of mining companies that are typically smaller companies with greater price potential. Speculating is also done in futures and options. (See Chapters 14 and 15.)

Understanding yourself first

Saving, investing, trading, and speculating aren't just limited to understanding precious metals. They're also to a great extent understanding yourself. Sometimes I get asked if an investment vehicle is appropriate and I have to provide a cop-out answer such as, "It depends." Any investment could be good but it depends on who the person is. How well do you understand yourself? Here are some points to consider:

- What are your goals? Are they short-term or long-term goals? Do you want to build wealth to buy real estate property next year or a large nest next decade?

- What is your profile? Are you just out of college or are you retired? Are you heading into your peak earning years or are you slowing in your pre-retirement years?

- What is your risk tolerance? Do you have the stomach to watch your investments rise and fall? Are you the type that sweats bullets when your nickel gets lost in the deep recesses of your couch?

- What is your investing style? Are you conservative or aggressive? Do you quietly sip your drink at the party or do you dance on the table with a lampshade on your head?

You need to understand what type of wealth builder you are or you will have difficulty. Personalities and investing vehicles do have a correlation. Half the battle in successful wealth-building and/or wealth preservation is knowing yourself.

The saver

The saver is actually not worried about a return on his/her money (like the investor, for example). The saver is more concerned about minimal risk and having money for a rainy day. The saver will have money in an account where the doesn't zig or zag; it just slowly grinds higher with compounding interest, ready to be used later for reasons both good (to finally invest it) or bad (pay for an unexpected problem such as a medical emergency or covering living expenses in the wake of a job layoff).

The investor

Are you an investor? Every investor is marked by the phrase "long-term" and this is indeed important. The classic investor does his or her homework to choose investments that are undervalued and that show great potential. The reason that these investors are "long-term" is that it takes a while for the market (other investors) to notice what you have and then buy it. If you have chosen your asset correctly, then "buy and hold" is your style and it has worked very well.

When people think of a successful investor the first name that comes to mind is invariably billionaire Warren Buffet. I can fairly label his investment style along the lines of "measure twice and cut once." He spends a lot of time analyzing the investment and if it truly measures up, he will buy it and hold it for a looooong time. I will tell you something interesting about Mr. Buffet that I recently told my clients.

Warren Buffet is not a spectacular investor but long-term you can't argue with the results. I told my clients that Mr. Buffet, to my knowledge, has never been the "top-ranked investor" in any year during his entire investing career. In other words, find his best year of stock-picking and there will be someone who did better in that given year. I doubt that he's ever made anybody's "top ten investors for the year xxxx.". And yes . . . he's had some bad or mediocre years. So what's his secret?

Warren Buffet certainly does his research and then invests but investing does take patience. He takes the phrase "long-term" very seriously. If the investment is growing, why get rid of it? He minimizes the transactions costs (fees, commissions, taxes, and so on), and he holds on to his winning positions indefinitely. In today's world, investors are impatient and are in a hurry to make a profit. And if they make it they sell too quickly to lock in their gain and then scurry off to hopefully make another choice. In other words, investors are frequently their own worst enemies. These principles and observations apply to precious metals as well.

The trader

For many, the words "trader" and "speculator" are one and the same. Trading is basically short-term in nature and more active in its scope than speculating. Markets do indeed go up and down constantly, and the ups and downs happen in the course of a year, a month, a week, or a single day. A market can have some amazing price swings that can occur in a single trading session. I've seen people buy, say, an option that morning and sell it in the afternoon and make a few hundred dollars (or more) in just a few hours.

Trading is usually about making money with the ripples in the market (the short-term ups and downs) versus riding the overall tide for perhaps greater profits over a longer term. They're happier, I presume, with lots of smaller profits versus waiting it out a few more weeks or months for bigger profits.

In Chapter 16, you find out more details on the different types of traders and trading styles. You may have heard about day-trading and swing-trading since they have had their moments of popularity in recent years. So trading can be profitable and rewarding, but to me it can be a pain: Trading can require a lot of attention; trades can go against you; then, of course, there are the transaction costs. Day-trading simply presents too many pitfalls for inexperienced folks. Besides, can you imagine me trying to day-trade while I'm watching my five-year-old and my six-year-old? Fortunately, I work on the first floor and the sharp objects are locked away.

The speculator

I love speculating! It's not for everybody, but it can be done with a small portion of your portfolio (again, if you don't mind the risks). The speculator does a lot of research and tries to position the speculative portion of the portfolio for a spectacular gain. You do the research and apply some rigorous economic logic and you can tip the scales in your favor. Let me use a great example of someone who did this with oil futures (though you can easily do the same with precious metals).

I'll call this speculator Joe. Joe opened a commodities brokerage account in the same firm that I use. Being bullish on oil, he opened the account in February 2004 with $90,000 and he concentrated entirely on call options on oil futures. A *call option* is basically a bullish bet that the asset (in this case oil) will be going up in price. Those of you who remember that year are already salivating — 2004 was a very bullish year for oil as it nearly doubled. Of course, investments tied directly to oil did well. In Joe's case, he used long-dated call options that he *rolled forward*. In other words, he would buy the options and sell them at a profit when they rose in value and then use the proceeds to buy new options with longer time values and so on.

The key to powerful profits is leverage and that's what you get with options. A single option can help you profit from the price swing of an oil futures contract that is tied to 1,000 barrels of oil. Every penny move in oil equates to a $10 move in the futures contract which in turn means a magnified move in the value of the option. If you didn't follow that don't worry: Chapter 14 covers futures and Chapter 15 tackles options.

What's the bottom line for Joe? He was right about the direction of oil and his original $90,000 account value became $1.1 million by December 2004. This does not include the fact that he took out $300,000 from the account! I guess he needed it to buy gasoline. Anyway, that's a great example of speculating done successfully. Just understand that for every successful "Joe" in futures and options, there are many unsuccessful ones. "Leverage" has made some speculators rich while it's sent many speculators into new professions where they say things like "Did you want that order to go?"

Knowing Whether to Get Physical or Own the Paper

First of all, let me make this very clear: I think that everyone should own some physical gold and/or silver. If you had to choose between the two of them, I like silver a tad better. In terms of currency problems, we are living in historic times. Currencies are losing value, which will only get worse as the major nations of the world keep their respective money supplies growing and growing. Using precious metals as an inflation hedge or counterbalance to the shrinking value of the dollar will become imperative. Whether you stick to the physical metal or own it on paper depends on your trading style, as the next sections point out.

In the following sections, if I don't specifically label something "physical" then consider it "paper" such as stocks, mutual funds, and similar vehicles.

For the conservative investor

For conservative investors and other folks who are risk averse, the precious metals area can be an unusual and skittish place. That's understandable, but they still deserve a place in your portfolio to some extent. Current and future economic conditions warrant it. If you're a conservative investor, consider the following investment vehicles:

- Physical: Bullion coins and/or bars (see Chapter 10)
- Large, established mining stocks (see Chapter 12)
- Mutual funds that specialize in precious metals (see Chapter 13)

The above may be conservative but that doesn't necessarily mean unexciting. Those who made it a regular practice to buy, say, some gold eagles (bullion coins issued by the U.S. Mint) every now and then during 2000–2006 would have been handsomely rewarded as the price of gold rose from $288 at the start of 2000 to a 2006 year-end price of $636 (a rise of 121%). Cool!

A good example of a large, established mining stock is Newmont Mining Corporation (symbol: NEM). It started the decade at $24 per share and ended 2006 at $45.15. Over a span of seven years NEM gained 88% not including dividends. Keep in mind that I'm using NEM as an example, not a recommendation. Not a spectacular performance but it certainly did better than the major market averages during that same time frame. The Dow Jones Industrial Average, for example, went from $11,358 at the start of 2000 to $12,463 at the end of 2006 for a percentage gain of almost 10%. Don't even think about asking about how Nasdaq did during that time.

Mutual funds traditionally have been a major part of investors' portfolios. For conservative investors, mutual funds are a great way to participate in precious metals. It's probably not a surprise to you by now; precious metals mutual funds (along with other mutual funds that concentrate on the natural resource sector) were definitely among the top-performing funds during 2000–2006.

For the growth investor

As world populations grow along with the money supply, there will be strong and growing demand for natural resources, such as precious metals. In other words, precious metals and their related investments are a very appropriate choice for investors seeking growth. With that, the following considerations are appropriate:

- ✔ Physical: Bullion coins and/or bars (see Chapter 10)
- ✔ A cross section of both small, mid-size and large mining stocks (see Chapter 12)
- ✔ Precious metals ETFs (see Chapter 13)
- ✔ Long-dated options on mining stocks and ETFs (see Chapter 15).

By the way, precious metals can be purchased directly for retirement in an Individual Retirement Account (IRA), which I cover in Chapter 10. In terms of acceptance and choices, precious metals have come a long way since the days the government banned private ownership.

A good example of a gold mining company appropriate for growth investors is Goldcorp Inc. (symbol: GG). It started the decade as a mid-size company priced at a very reasonable stock price of $2.59 (price adjusted for stock split). GG ended 2006 at $28.37. The percentage gain would have been over 995% (not including dividends). That's more like it! Sure the stock price hit bumps and potholes along the way, but patient growth investors reaped an excellent gain.

Now options can be a dicey matter. Some call them vehicles that can be part of the growth portfolio while a larger crowd would label them speculative without question. The honest answer is "it depends." Did I shock you? Options are very versatile and they can be applied in various strategies of differing risk levels. There are even some options strategies that are appropriate for conservative investors and for income-oriented investors. For those buying options, be they call options or put options check out Chapter 15.

For the speculator

In recent years, I have had the pleasure of seeing some of my clients (who didn't mind the added risk exposure) see $40,000 in risk capital (money, if lost, that didn't force you to eat dog food) and turn it into $250,000 in less than three years. Speculating, again, is like financial gambling, but I don't do it justice calling it that because enough knowledge and research can greatly offset the "gambling" portion of it. There is nothing wrong with speculating (heck, I do a lot of it), but it does require more diligence, research, and risk tolerance than other styles investing. The "thrill with your money" can definitely turn into the "kill of your money," and it can happen all too suddenly.

With all that said, here are good considerations for speculators:

- ✔ Small mining stocks
- ✔ Options on mining stocks
- ✔ Precious metals futures
- ✔ Options on precious metals futures

My favorite of the four choices is options but hey . . . all the choices I list are good. In addition, even speculators should get some physical gold and/or silver as well because inflation will affect speculators, too.

For the trader

Even people who trade also have some savings and investments elsewhere in their financial picture. To me, trading is an active pursuit whereas the other areas are technically considered passive. I even consider speculating passive because it's a case of having your money work for you in spite of the added risk. Speculating could be long-term as well as short-term, but trading is exclusively a short-term pursuit. Therefore, trading best occurs with assets in a market that can move far and fast. Also, successful trading tries to create or find profit opportunities in both up and down moves in the markets.

Therefore, the best venues for trading are:

- ✔ Precious metals futures
- ✔ Options on mining stocks
- ✔ Options on precious metals futures

I cover more on trading in Chapter 16.

Getting the Amount Just Right

I can't give you any hard and fast rules for what percentage of metals should be in your overall financial picture. I can't turn to you and say, "Yup . . . 11.3% is a must for you." I know guys in their 40s who will have a fairly small portion of their assets in precious metals, and I know some senior citizens in their 70s with large holdings. Some guidelines you just can't find inside this or any book — they're inside of you. You're an individual, and what's a right amount for you may not be for others and vice versa. Remember that I am writing this in 2007. I definitely would not have written the same guidelines in 1997 and I have no idea what I'm going to write in 2017 but now . . . today . . . precious metals are a necessity. I don't see that changing for the rest of this decade. With that said, the following sections give you my take on what amounts each different type of investor should shoot for.

For the conservative investor

I think that conservative investors ought to consider having 10% of their financial portfolio in the physical metal (coins, and so on) and 15% in paper such as mining stocks. This should span both the precious metals such as gold, silver, and platinum along with base metals and uranium. The emphasis should be on larger, established companies in politically friendly countries. Of course, if you do more or a little less that's fine since your personal style and comfort level come into play as well.

Although this book is intended for precious metals let me say (as a public service going beyond the call of my author duties) that some of your other portfolio holdings ought to be in areas such as consumer staples, energy, and other necessities. Take it up with your financial advisor and tell 'em I sent ya.

For the growth investor

Just as with the conservative investor, I like the 10% in physical and the 15% in paper approach. Of course, the character of the 15% paper portion can be different and certainly more aggressive. Look into junior stocks and mid-size companies. Learn about options and how they can boost your portfolio.

For the speculator

After you get past the 10% in physical, consider 20% in the same type of venues as the growth investor and 10% in the speculative area, maybe more if you feel you're up to it and you have enough diversification in the rest of your financial situation. The 10% (or more) speculative portion can be options in a stock brokerage account or futures and/or options in a commodities brokerage account.

For successful speculating, I believe that you shouldn't try to spread that speculative 10% portion around in a batch of metals. I know that I am speaking heresy in some circles but hear (or read) me out. Successful speculating requires time, research, and attention. Watching half a dozen different speculative vehicles makes it difficult. It reminds me of that guy I saw on the Ed Sullivan Show (yup . . . I'm that old) who would have a bunch of plates spinning on sticks and he would keep running to each one while his lovely assistant smiles from a distance . . . close to the nearest broom.

In the past few years I traded silver options almost exclusively, and I got to know the silver market very intimately. It paid off since the average account gain for my clients and dedicated students for three consecutive years was about 100% per year on an annualized return basis. It was very gratifying to accomplish that. The broker servicing the accounts at the commodities brokerage firm tells me that this is very difficult to do and is a higher success rate than most of the programs in the industry. For me the secret was that speculating is best done like a laser beam and not like a shotgun blast.

Before you take the plunge, test the water first. Do some simulated transactions. Web sites such as www.Marketocracy.com and www.stocktrak.com have great programs for simulated or paper trading without risking your money.

For the trader

This is a different animal. I hope that traders do put 10% into physical and maintain a diversified portfolio not unlike the growth investor. After that they can take the risk capital portion of their portfolio much like the speculator and start trading. If you are considering trading as a part-time activity, then you may want to consider the simulated trading programs that can be found on the Internet to test your approach before you decide to risk your money (see above section for suggestions).

As in speculating, trading is best done if you watch a single market like a hawk. Because the transactions are more short-term in nature and you're catching the ripples, it is important to have a plan in place. The first thing you should plan to do is simulated trading (see above section for suggestions). Next, instead of trading futures directly, do options so that you greatly limit your risk. Options have great leverage but you can also limit your risk going in. Lastly, plan to read up on the trading styles of successful traders. One of the greats in the history of financial markets is Jesse Livermore. His thoughts on trading can be found in the book *Reminiscences of a Stock Market Operator* (John Wiley & Sons). Just a thought . . . speaking of which, I have many more thoughts on trading in Chapter 16.

Chapter 3

The Beauty and Benefits of Metals

*I*n 2006, I celebrated my 25th year in business. Looking at investments was an uneventful moment outside of looking at their financials and so on because investments are generally intangibles; you can't touch or taste them (not that you would want to munch on a stock certificate), but they are more or less abstract things. Precious metals — and I'm talking primarily gold and silver — can be beautiful things that you can actually hold and admire. Once in a long while I get the chance to look at the gold eagles and the silver eagles that I have acquired over the years, and I see that above and beyond the fact that I got them for cold, hard economic reasons I am reminded that physical gold and silver can also be works of art.

Fortunately, the beauty can also extend to your wealth-building picture as well. This chapter covers the benefits of precious metals and their related investments. Don't think that I am mesmerized though; Chapter 4 goes into the ugly side of metals (risk).

Protecting Your Portfolio Against Inflation

Inflation is one of the greatest dangers to your investing and wealth-building pursuits. Many people make the common error of thinking of inflation in terms of its symptoms instead of seeing inflation for what it really is, and I'm not immune to such an error. I've been known to think, "Gee whiz . . . how much for that bag of groceries?! Here . . . why not take my whole freakin' wallet!! Never mind . . . I'll just put back the breath mints. Harrumph."

But high prices aren't the cause of inflation; they are the result or symptom. Inflation means that the government is printing a lot of money (dollars or euros or whatever the currency is of that particular country) and this increasing supply of money is used to buy a limited basket of goods and services. The following sections show you why diversifying your portfolio with precious metals can strengthen your wealth building.

Precious metals against the dollar

Monetary inflation refers to increasing the money supply ("monetary" is just an official way of referring to "money" such as dollars). Monetary inflation then leads to price inflation, which is what most people are familiar with. The bottom line is that you end up paying more for the stuff you regularly buy. The end result isn't that goods and services cost more — the money you use is worth less. That's the insidious side of inflation; it acts like a hidden tax that destroys your purchasing power. Take a look at the following charts. Figure 3-1 shows you the value of the U.S. dollar since the beginning of the decade. Now compare that with Figure 3-2, the chart of gold's performance, and Figure 3-3, the chart of silver's performance. You tell me which is the good and the bad (before I get ugly):

Keep in mind that whatever investments you have, they are denominated in dollars. If you are a U.S. citizen then the odds are pretty good that virtually everything you own is measured and valued in American dollars. Therefore, as the dollar goes down in value, that will have an adverse effect on your financial well-being.

If you had $1,000 tucked away in your sock drawer in 2006 and you checked in 2007 and it's still there, you might say that you still have $1,000 (or "Thank God no one checks this drawer!"). From a nominal point of view, your wealth hasn't budged. It's still $1,000 and you feel fine. But before you check your mattress, understand that your wealth actually shrunk. The official rate of inflation for 2006 was 2.8%. That means that, according to the U.S. Department of Commerce (more specifically, a subdivision called the Bureau of Labor Statistics found at www.bls.gov), you would have needed an additional $28 in your sock drawer just to buy the same "basket of goods" as the original $1,000 from a year ago.

Now before you extract $28 from your mattress and transfer it to your sock drawer, keep in mind that the "official rate of inflation" is probably not accurate. The fine print for the official rate of inflation states that it does not include food and energy costs. Yes . . . I find that aggravating, too. According to the Bureau of Labor Statistics, as long as you don't eat food, use a car, or turn on lights in your home . . . you're doing fine! For 2006, had you calculated the real-world rate of inflation, you'd be looking at a rate somewhere in the 6% to 9% range. Fortunately there are alternate sources of inflation data and analysis available at places such as www.inflationdata.com and www.shadowstats.com to help you get a more accurate assessment of the inflation problem. The reality for your sock drawer is that you really should have an additional $60 to $90 added. Hmmm . . . time to replenish your mattress.

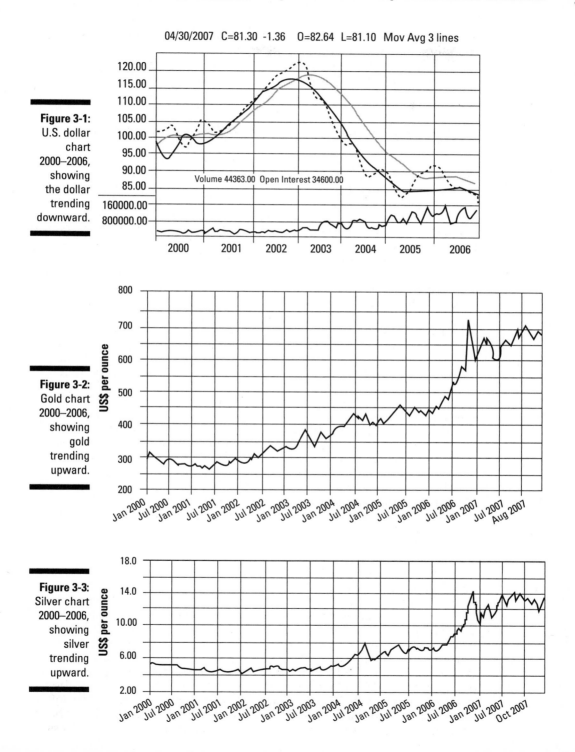

04/30/2007 C=81.30 -1.36   O=82.64   L=81.10   Mov Avg 3 lines

Volume 44363.00 Open Interest 34600.00

Figure 3-1:
U.S. dollar chart 2000–2006, showing the dollar trending downward.

Figure 3-2:
Gold chart 2000–2006, showing gold trending upward.

Figure 3-3:
Silver chart 2000–2006, showing silver trending upward.

Inflation is not a short-term problem. In the above example, whether the actual difference is $28 or $60 or $90 is not that big a deal. But . . . it becomes a very big deal over an extended period of time. In 2007, the U.S. dollar has lost over half its value from 20 years ago. That's a problem and that's why diversifying into tangible assets such as precious metals makes sense.

Diversification against all currencies

You encounter a purchasing-power risk in any manmade currency that could be produced at will by a government authority (such as a government central bank). What does this mean for you going forward? As I mention in other chapters, a growing (mismanaged?) money supply is at the heart of the problem. Because paper or *fiat* currencies first started being used centuries ago, governments have never resisted the temptation to print up lots of it. This is why the dustbin of history is top heavy in currencies that were over-printed into oblivion. Could that be the case today? Are currencies being created in ever increasing amounts?

Oh yeah. Inflation isn't just a problem for you and me. It's affecting many people across the globe. Gold and silver are doing well and should continue to do well regardless if we are talking dollars, euros, francs, yen, or salt. It is safe to say that hyper-inflation, a currency crisis, or collapse is destined to hit in the dollar and/or other currencies at some point in the coming years.

What does this have to do with the benefits of precious metals? It's like looking through a window, watching a torrential downpour, and discussing the benefits of umbrellas. Yes . . . inflation will rain on your wealth-building parade and the forecast is for a batch of storms on our radar screens for the foreseeable future.

The most obvious and familiar precious metals are gold and silver, and these two should suffice for most portfolios. However, precious metals also encompass the following . . .

- **Platinum:** Although you can get platinum in physical form such as bullion coins, it is not a huge, liquid market. For more details see Chapter 7.

- **Palladium:** A cousin to platinum that is also not a huge, liquid market (see Chapter 7).

- **Uranium:** Not meant for your sock drawer (see Chapter 8).

- **Obscure precious metals (such as rhodium):** Too obscure for this book in today's marketplace . . . perhaps a *Rhodium For Dummies* . . . in the future.

I also touch on base metals in Chapter 9 because they do deserve a spot in some portfolios. In recent years, some base metals have seen their prices double and triple. Those gains would qualify as "precious" in my book (at least this book). Even though copper and nickel are readily found in the composition of your pocket change, they are base metals that offer investment opportunities as well.

Benefits for Investors

You always need a good reason to do something, especially when you're talking about where to put your hard-earned, inflation-ravaged money. I think that precious metals in their various formats offer many benefits (see the following sections), and the sum total of these benefits makes them a compelling choice as a part of your portfolio.

Safe haven

One point that you will hear on a regular basis from gold aficionados is that gold and silver are real money that are someone else's liabilities. For newcomers to the world of precious metals that would be an intriguing statement. What does it mean?

A currency such as the U.S. dollar is backed up by the "full faith and credit of the United States." You can surmise that a dollar is not money in the truest sense of the word; it is a type of claim or liability backed up by the issuing authority which is in this case the federal government. There was a time when the U.S. dollar was redeemable in gold and silver. That dollar was backed up by something that was universally considered a "unit of exchange" and a "store of value," which is what gold and silver were considered for literally thousands of years. This was the whole point behind the "gold standard."

As long as the dollar was backed by gold, it had strength and people had confidence in it. Starting in 1913 with the creation of the Federal Reserve (America's "government central bank"), the dollar started to decouple from gold until the connection between gold and silver was totally severed during the Great Depression. At that point, the spigots on producing dollars opened up. The U.S. dollar started a long, slow-motion bear market where a dollar in 1913 is only worth two cents today. In all that time, the dollar lost 98% of its value!

Precious metals such as gold were a different story. It has been said that a single ounce of gold could have bought you a decent suit during the 1800s. If you sold a single ounce of gold on December 24, 2006, you would have gotten $620 (not including transaction fees) and that would have been plenty of dollars to buy a nice suit for Christmas.

In good times and bad, precious metals have retained their value and earned their title as a safe haven.

Privacy

In the world of financial assets, reporting such as 1099s is common. With certain types of precious metals vehicles, there is no reporting (for transactions under $10,000 value). This is especially true of gold and silver collectibles such as numismatic coins. Of course this does not relieve you of the obligation of paying taxes in the case of capital gains transactions.

Rules can change obviously especially in post-9/11 times so make sure that you discuss your purchases and sales with your tax advisor.

Inflation hedge

In early 2006, Harry Browne died. I thought it was a great loss to the investment and political world. (He ran for President on the Libertarian platform in 1996 and 2000.) His claim to fame is that he is a legendary investor who was right on the money during the 1960s and '70s. One of the last books that he wrote was a short book that was a concise, true gem of financial wisdom called *Fail-Safe Investing: Lifelong Financial Security in 30 Minutes* (St. Martin's Press). In it he detailed a model portfolio that was easily constructed by the average investor and it had performed very well in a variety of economic conditions. It consisted of 25% each of cash, stocks, bonds, and . . . gold. He suggested that you rebalance it each year to keep the 25% allocation.

Mr. Browne was keenly aware of the dangers of inflation, recession, and other systemic problems that occur because of political and governmental mismanagement (such as through inflation, taxes, and regulation). He recognized that gold was not easily produced and manipulated by government. One question that may pop up is "Can't gold be produced as well?" Gold is mined but the average annual addition of gold to our world gold supply is roughly 2%. That physical limitation keeps gold growing at a pace that helps it to retain its value yet plentiful enough to keep pace with population growth.

Dollar hedge

A dollar hedge seems synonymous with an inflation hedge since you're dealing with the same issue. But *dollar hedge* is more related to those who see the dollar as an entity worth trading against. You can invest or speculate based on the direction of the U.S. dollar as its own investment vehicle.

In the same way you can look at the relationship of a common stock to the company it is attached to, you can look at a currency as it is attached to a country. There are definite similarities. When a country's currency is strong, it is to the rest of the world a representation of how strong or well-managed the country is. During the 1980s and 1990s, the dollar was considered a strong currency, but in this decade, due to factors including a ballooning national debt and a large trade deficit, the dollar started a long decline. Coupled with the fact that we keep printing more and more of it, that trend will probably continue.

A hedge against the dollar can manifest itself in a variety of ways. For example, some will speculate in the currencies of other countries because if one currency declines in value then some other currency is probably rising. Of course, the other currencies could also decline as well. The bottom line is that gold and silver are solid choices as dollar hedges.

Confiscation protection

In Chapter 4 I spend some time dealing with the risks of precious metals and in one segment the focus is on political risk. The issue was how government can affect your investment. In the 1930s, the FDR administration authorized government confiscation of gold and essentially banned ownership of gold. But they exempted gold coins, having a recognized special value to collectors of rare and unusual coins.

Even in the extreme case of heavy-handed government intervention, gold can still be a store of value.

Liquidity

Liquidity is an important benefit for investors. All *liquidity* really means is how quickly you convert an asset to cash. Securities such as stocks and bonds are very liquid assets. Assets such as real estate are not. Precious metals and their related investments tend to be very liquid. Mining stocks, like other stocks, can easily be sold in seconds either with a call to the broker or a visit to a Web site. You can do likewise with options and futures (you get them and get rid of them through your brokerage account).

In terms of physical metals, the most liquid precious metals are gold and silver. If you have gold and/or silver coins, they are very liquid as you can go to virtually any coin dealer store or Web site and quickly convert bullion or numismatic coins to cash.

Numismatic coins carry a high markup so if you were to liquidate them you may get only 70 cents on the dollar (or less). Most people sell their coins to coin dealers but these dealers will basically offer you wholesale prices since they will end up trying to resell the coins at retail. Consider selling coins to other investors or collectors as they are easier to find now through the Internet.

Portfolio diversification

In the book *Stock Investing for Dummies*, 2nd Edition (John Wiley & Sons), I wrote extensively on portfolio diversification. It was important to have stocks in different sectors so that the overall portfolio could be safer yet still grow without you having to reach for the antacid on those volatile days. In this book, I get to tell you what I think is a good part of the diversification. Metals mining stocks (and uranium) have been among the top-performing stocks in the country in this decade. Yet, I am sure that many in the public haven't realized it. You already know about physical metals and their performance along with their advantages. Their "disadvantages" are covered in Chapter 4. How about the stocks?

Take Agnico-Eagle Mines Ltd. (symbol: AEM) for example. AEM is a Canadian gold-mining stock listed on the New York Stock Exchange. You could have purchased it at the beginning of the decade for around $5 a share. It hit $45 in 2006 for a total long-term gain of 800% (not including dividends). If that's diversification, I want more of it. Other selective mining stocks in gold as well as other metals (and uranium) had similar results. I realize that the key word is "selective," but you can find more details on choosing mining stocks in Chapter 12. Yes, I spill my stock-picking guts there!

Your retort to me after that paragraph might be, "Gee, Paul, I'm sure that if I look hard enough that I will certainly find stocks that did just as well and probably a heck of a lot better in totally different sectors. Why choose stocks in the metals sector if there are also healthcare, biotech, and so on?" I'm glad you asked. Or I'm glad that I wrote down that you asked what I'm glad you asked.

In the realm of investing and speculating, the past is something you learn from so that you can prepare for the future. In that prior example, AEM went up 800% in the same time frame that the asset it mined (primarily gold) went from under $300 to over $650 an ounce (gold more than doubled). I, along with an army of gold experts, see economic conditions that are very fertile for gold to shoot past $1,000 and much higher in the coming years (more exciting details on gold in Chapter 5). In the event that gold does go into four figures, what happens to gold-mining companies that are profitably sitting on millions of ounces of the stuff? Can you say "jackpot"?

It doesn't stop there. Consider the rest of the metals that are covered inside the pages of this book. Gold is great but there are also silver, uranium, and so much more. Cool profits in hot markets; I like the sound of that.

Benefits for Traders and Speculators

Traders and speculators want a market that makes profits easier to achieve. The current and pending precious metals market offers an environment for just that.

In the next sections, I want you to consider supply and demand as the ultimate factors in your wealth-building pursuits.

Supply and demand

As a trader or a speculator, what do you think of when you hear the phrase "supply and demand"? Say what? Do you think of Mrs. Krapalbee in your college econ course on the first day of the semester? Let me change that dynamic. Think "profits." When you understand this simple, age-old, foundational economic idea and apply its eternal wisdom to today's fast-moving, modern markets, you can make a profit easier. This idea sits at the heart of your trading/speculating success.

The natural resource (or commodities) sector is wonderfully positioned for gains in the coming years. Breaking it down a step further, natural resources can be neatly separated into two segments: finite natural resources and renewable natural resources. *Renewable natural resources* include things such as corn, sugar, coffee, and so on. Now don't get me wrong — I like these areas and they have been profitable markets since natural resources have been in a multiyear bull market and it will continue for years to come.

But you know what happens with renewable resources. The news reports "the coffee crop in Brazil will be down this season" and the price of coffee goes up on the news and you say "time to percolate some profits!" and you buy coffee futures. Then . . . as you smell the aroma of fresh-brewed money . . . wham! The next report says "Chile expects record coffee harvest" and your coffee futures nose-dive as you say "I knew I shoulda done cocoa instead!" Well, I know that over-simplifies it and maybe I'm a little biased. These are great markets with great opportunities but I prefer to speculate with finite natural resources. Yes . . . by a country mile. Why?

Finite natural resources refer to resources that are not readily renewable or replaceable. Good examples are oil, natural gas, base metals, uranium, and precious metals. In other words, whatever we find that is above and below ground . . . that's it! No more. Our society ultimately finds substitutes, but only if it really, really, really has to. That's humanity. We go through a painful transition as we move from one resource to the next but we hang on to the original resource until a suitable substitute comes along. A good example is oil. Modern societies use oil and will keep using it until it's gone, but progress has been too slow into cleaner substitutes such as solar and wind energy. In the meanwhile, think about the following dynamic:

- The world population keeps growing.
- The nations of the world keep expanding their money supplies.
- Finite natural resources are slowly being used up.

Put those three together and you have dynamic opportunities in things such as . . . precious metals. The benefits for traders and speculators then become obvious. I cover trading strategies in Chapter 16.

Huge gains potential

I have one client who opened a commodities brokerage account in late 2003 with $40,000. Options on silver futures were purchased. After a roller-coaster ride complete with bone-jarring corrections, the account was worth over $300,000 by May 2006. A 650% return in about two and a half years. Nice! That doesn't include the fact that the original $40,000 was taken out of the account and sent back to her. Nicer! This was of course speculation but it was done with only 10 percent of her money. Silver in that time frame went from under $5 in 2003 to over $14 in May 2006.

What would happen to an account like that if silver was to go to its old high of $50? Of course, $50 was its old high in 1980. Adjusted for inflation it would be in three figures. That's the type of potential that makes speculators salivate.

Trading versus speculating

I know an associate who likes to speculate and he hates trading. Trading is too much work and it requires a lot of time, attention, and timing. Plus you add in the transactions costs. After you factor it all in, you may make a profit of . . . say a few hundred bucks. Now there are traders out there who make some great profits and even a good living at it. Trading with options and/or futures has a lot of advantages. As for me . . . I'm not a trader. I'm a speculator.

Being in a full-time business and helping to raise my two young boys means that trading is not right for me. Successful trading takes a lot if you are going to make it work. I prefer speculating because I can do all of my homework and make a few transactions such as buying that small stock with great potential or getting a long-dated option on, say, a silver futures contract and then wait for the fireworks. This has worked better for me.

The benefits of speculating

As I have mentioned before, speculating is akin to financial gambling, but if you do your homework, you can greatly increase the odds in your favor. It is possible to take a small amount of risk capital and parlay it into a much larger sum. Here are some benefits of speculating:

- ✔ **With knowledge, research and discipline it is possible to make great money.** During 2004–2006 virtually everyone of my long-term clients either made good money or great money. I know real-life examples of clients who turned $5,000 into $20,000 or more in a few months.

- ✔ **You can do it with little money.** I know several clients and students who started with as little as $1,000 in futures options and some who started with as little as $500 in stock speculating and all have made great profits.

- ✔ **You can minimize your risk.** Many think that to turn a small sum into a great sum that you need more leverage. Not really; you can utilize vehicles that have leverage with the risk of leverage (such as options).

Futures is a great area for speculating and I certainly cover futures (in Chapter 14), but my favorite for speculating is options. Whether it's options on stocks or on futures, my belief and experience tell me that it is a great way to profit without incurring undue risk. The phrase "undue risk" sounds loaded, but it will make sense. Futures carry risk beyond your purchase price. You could get into futures with, say, $5,000 and within days lose the entire sum plus still owe thousands. Yikes . . . who needs that?

This is why it is important that besides knowing the benefits of speculating you should understand the risks before you begin transacting. For those who want to speculate with unlimited profits while limiting risk and loss, consider options. I love options and I get into greater details on them in Chapter 15.

Chapter 4

Recognizing the Risks

*I*n my financial seminars I spend a lot of time talking about risk because it obviously needs to be dealt with, but also risk is entwined with the deepest desires of most investors . . . increasing return. It goes without saying (so I'll write it down) that the age-old equation in the world of investing is risk versus return. This equation states that if you want a greater return on your investment, then you have to tolerate greater risk. If you don't want greater risk, then you have to tolerate a lower rate of return. The world is full of pitfalls and precious metals are no different but keep in mind that precious metals can excel when your other investments don't. Precious metals guard or hedge against risks that can hurt conventional stocks, bonds, or other fixed-rate vehicles. Here are some of the risks against which precious metals excel:

✔ **Purchasing power risk:** As inflation rears its ugly head, this results in higher prices.

✔ **Currency crisis:** As nations increase their money supplies, the long-term result is usually a crisis or even collapse of the currency.

✔ **Geo-political risk:** This can range from war to terrorism to international strife.

✔ **Systemic financial risks:** When a crisis occurs due to problems with vehicles such as derivatives.

A good example of the value of, say, gold or silver is what happened in Zimbabwe in 2006. On paper almost everyone is a millionaire but most people are really poor! The country underwent hyper-inflation as the inflation rate hit the nose-bleed level of 914% in March 2006. Basic goods like a roll of toilet paper soared to $145,750 (only 69 cents in U.S. dollars). Gee, that might encourage some of their citizens to . . . uh . . . use Zimbabwe's currency instead. As incredible as that hyper-inflation was, it is actually not an odd example because it also recently occurred in Argentina and Serbia and numerous times across history. Those folks who had gold, silver, or other hard assets fared much better than those who didn't.

In this chapter, you not only get the chance to check out what types of risks are associated with precious metals investing, but you also find ways to minimize those risks.

What Risk Means to You

Before I make you paranoid about risk, keep in mind that it is ubiquitous and just a normal part not only in building wealth but also in living life. Heck, just getting out of bed in the morning could pose a problem. Here are the various types of risks (now that you are out of bed):

Physical risk

Now I don't mean that you may hurt your back from picking up precious metals (heavy metal is a different animal altogether). It just means that if you have gold in physical form then you have to understand that having it has risks as does owning any valuable property. You have to keep it safe. For some that means keeping your physical metal (such as gold, silver, and platinum) possessions in a safe-deposit box at the bank. For others it means in a secure hiding place at home.

You have to decide. Gold as a physical holding means you need to be concerned about the risk of loss or theft. Then there's relative risk; in other words, what if your relative finds out? Removing some risk always means common sense. After all, if gold hits $3,000 an ounce, I don't think you should be boasting about your investing skills at the local bar.

Market risk

Market risk may be the most prevalent risk associated with gold. *Market risk* refers to the fact that whenever you buy an asset (physical, common stock, and so on) its price is subject to the ups and downs of the marketplace. In gold, as in many investments, the price can fluctuate and it could do so very significantly. What if you buy today but tomorrow there are more sellers than buyers in the gold market? Then obviously the price of gold would go down. The essence of market risk in commodities such as precious metals is supply and demand.

Another element of market risk can occur when you are involved in a *thinly traded market* — in other words, there may not be that many buyers and sellers involved. This is also called *liquidity risk.* This can happen, for example, in futures. Although futures are usually a liquid market (an adequate pool of buyers and sellers) there may be some aspects of it when it might not be that liquid. Say that you want to sell a futures contract that you recently bought that is not an actively traded contract. What if there are no buyers when you want to sell? Your order to sell through the broker may sit there for a long time. The sale price of the contract would drop and you would lose some gain or even end up with a loss. Be sure to communicate with the broker regarding how active that particular market is.

It is probably appropriate to place in this segment the market risk of mining stocks. The stock of mining companies certainly can go up and down like most any other publicly traded stock. Stock investors can sell stock when they see or expect problems with the company. If, for example, you are considering a gold-mining company, the risk to consider is more that just the fact that it is into gold and the commensurate market risks with gold itself. Also consider the company. Is management doing a good job? Is the company profitable? Are sales increasing? How about their earnings? Do they have too much debt . . . and so on. Mining stocks are covered in Chapter 12.

Exchange risk

This one sounds odd. What the heck is exchange risk? Well, it's not a reference to currency exchange; *exchange risk* is a reference to the risks that could occur at the exchanges where futures and options are traded. When futures and options are transacted at the exchange such as at the Chicago Board of Trade (CBOT) or the New York Mercantile Exchange (NYMEX), they are done so under the rules and regulations of the exchange. The exchange can either purposely or accidentally encourage market outcomes by changing the rules and regulations on an ongoing basis.

A real-life example happened in the Spring of 2006 with silver futures at the NYMEX. Silver was rallying nicely as speculators were buying into silver futures contracts. There were great expectations for silver due to the then-pending silver exchange-traded fund (ETF). (You can find out more information on precious metals ETFs in Chapter 13.) The exchange decided to raise the margin requirements on margin contracts to try to quell over-speculation. When normally you could put down 10% to speculate on a futures contract, the new amount would be raised to, say, 12% or 15% or possibly more. When you require people to put more funds in for the ability to speculate then of course you will diminish that activity. If the margin requirements are raised too high that will result in more selling. More selling results in prices dropping.

The exchanges want an orderly market and they may change regulations or adjust requirements to encourage or exact an outcome. Sometimes that outcome may result either purposely or accidentally in a negative way for you. Here are the three events that may happen at an exchange:

- **Changing margin requirements:** I gave an example above and this is the most common event that an exchange could enact.

- **Liquidation only:** This is a rare event but it can happen. This means that the exchange may temporarily restrict the buying side and only selling can occur thus forcing the price down. This occurred with silver in 1980 when it hit its then all-time high of $50.

- **Halt trading.** Another rare event. The exchange may temporarily halt trading in a particular futures contract.

Political risks

Political risk is probably one of the biggest dangers that investors and speculators don't see coming. It is the one that comes out of the blue and blindsides your portfolio. *Political* refers to politicians who in turn run government. As far as we are concerned, political risk and governmental risk can be synonymous. In my seminars I mention that politicians are Dr. Frankenstein while government is Frankenstein's monster. The bottom line is that *political risk* means that government can change laws and regulations in a way that can harm your investment or financial strategy. This can happen in your own country or by another country. Consider what happened in the 1930s right here in America.

In 1934, FDR and congress passed the Gold Reserve Act which made gold ownership illegal. Had you bought gold in prior years to preserve your wealth in the midst of the Great Depression, well . . . you were now out of luck. FDR then issued a presidential order fixing the price of gold at $35 an ounce which stuck for decades to come. FDR didn't want private citizens to have an alternative outside of the official paper currency.

Fast forward to our times. Political risk is alive and well (unfortunately). In many countries such as Bolivia and Venezuela, the government nationalized properties (government taking private property by force) by foreign companies, among them, mining companies. Had you owned stock in these mining companies you would have seen the share prices drop. Sometimes the share prices drop at the mere threat of government action. In 2005, for example, the Venezuelan government mentioned that it may take property owned by the Toronto-based gold-mining company, Crystallex International. Its share price fell by a whopping 50% in a single day. Venezuela's dictator Chavez did increase taxes on many foreign companies while nationalizing some industries (maybe he needed the funds to buy toilet paper from Zimbabwe).

That's the problem with political risk. As an investor or speculator, you could do all your homework and make a great decision with your portfolio backed up by great research and unflinching economic logic and still lose money because of a government action that could have been unforeseen.

An ounce of prevention is worth a pound of cure. It is best to stay away from investments (such as mining companies) that are too exposed to risk in a politically unstable or unfriendly nation. There are still plenty of precious metals opportunities in politically friendly environments such as the United States, Canada, Australia, and Mexico (at least until the next election!).

The risk of fraud

The risk of fraud is as real in precious metals as in every other human endeavor. It's tough enough trying to make a buck when the market seems honest. But we must understand that as a market becomes popular or "hot" so it becomes a target for scam artists. Fraud can materialize in a variety of ways but I think that it can be safely categorized into three segments, which I discuss in the following sections: scams, misrepresentations, and market manipulation.

Scams

Scams are those events that the consumer organizations always warn about. The image is conjured up about those boiler-room operations where a slick con artist calls up a little old lady in Pasadena and talks to her about riches to be made in gold and silver if she could crack open her piggy bank and send off a nice money order chunk of her savings. This is certainly a real risk and it becomes more apparent when the source of potential fraud is popular. When Internet auctions became a hot consumer area, there were more Internet auction–related scams. When the real estate market became red-hot in 2005, there were more real estate scams. When precious metals become the "bubble du jour" then we will need to be wary of scammers here as well.

Misrepresentations

I put this as a separate topic because it can be a different animal. Basically the point is that you may put your money into a venue and you may not be getting what you think you are buying. A good example is what the respected silver analyst Ted Butler recently warned about regarding silver certificates. There have been millions of silver certificates issued in recent decades but there is the real possibility that there isn't any real silver backing them up. In other words there are purchasers of silver certificates who believed that they could convert their paper into actual silver in due time but in fact will not be able to. That sounds like misrepresentation to me.

Market manipulation

Earlier in 2007, the financial media reported a serious matter regarding naked short selling among the smaller stocks. Short selling can send the stock's price plummeting. Some large brokers and their clients have been caught illegally profiting through a manipulative technique called naked short selling. This was an especially egregious activity with the stock of smaller mining companies. In *naked short selling,* the perpetrator can sell massive quantities of stock essentially created out of thin air to force the price of the stock to come crashing down. Imagine if you owned shares of a small mining company and you saw the share price plummet by 40% or 50% or more for no apparent reason.

Minimizing Your Risk

Including precious metals (to whatever extent) in your portfolio minimizes risk because precious metals, such as gold and silver, have historically helped investors in times of economic uncertainty and political and financial tensions in the world at large. The long-term picture for precious metals should continue to bear this out. But as this chapter points out, nothing is without risk. Precious metals do carry risks and you can minimize your risks — just check out the following sections.

Gaining knowledge

I remember getting into options on silver futures some years ago. It was early 2004 and silver was rising very well and my account was performing superbly. I thought to myself "gee what a genius am I!" then along came April 2004. Silver plummeted by nearly 40%. I thought to myself "gee what a moron am I!" In retrospect, it worked out just fine and I made some great profits but I was sweating bullets that spring. Seeing the value of your "investment" drop by 40% can make you freak out. I would have jumped out the window but I'm on the first floor. The point is that I learned that precious metals futures and options can have very wide and scary price swings. That is the nature of the market.

In fact it is not uncommon for precious metals to correct by 20% to 30% or even more at least once a year (I've come to learn that a "correction" seems awfully incorrect at the time).

The terms *correct* or *correction* means that a market came back down after going up too far and/or too fast. Don't confuse a market correcting with a market experiencing a bear market. The correction is a temporary pull-back in the price of the asset that is in a long-term bull market or rising market. The term *bear market* means that the asset is in a long-term falling or decreasing market. In other words, the difference between a correction and a bear market is the same difference as fainting and dropping dead. In the former you recover and get back on track.

Being disciplined

When markets go up and down, it can be difficult to stay disciplined. Markets can't be controlled by individual investors, but investors can and should exercise self-control. People can let their emotions overrule their thinking and do the wrong things when they ought to do the opposite. It happens especially in fast-moving markets.

My client (I'll call him Bob) had put $2,000 in a commodities brokerage account. Under my guidance he purchased some options on silver futures. Within a few weeks the value shrunk to $900. Imagine that; he was down 55%. Of course he was very concerned (wouldn't you be?), and we discussed the situation. I told him to stay the course because my research told me that the underlying asset (silver) was in a bull market and that this drop in price was a temporary condition. In addition, the options that were purchased had two full years to go before they expired (options are covered in Chapter 15). This is an example of what happens in the marketplace; no matter how solid your research and logic, the market can go against you. If your research and logic were sound then the odds would swing back in your favor in due course. He decided not to panic and stay the course.

For Bob, the discipline paid off. The $2,000 he speculated with became $4,000 a few months later as the silver market rebounded. Those options were finally cashed in for $5,000 about 12 months after the initial purchase for a gain of 150%. To this day, the account is still growing as we stayed disciplined and bought (and cashed in) at points that made sense. Bob could have panicked when it was at $900, cashed out, and leaped out the nearest window, but thankfully he stayed disciplined and reaped some excellent profits.

 When you are in fast-moving markets, have a plan in place regarding how much you will put at risk, when you plan to get in, and under what conditions and price points you will take profits (or losses).

Being patient

Everyone wants to get rich quick. Who wouldn't like to make a fast buck? Well, people who reach for the fast gains end up with fast losses. I can say with confidence (verifiably) that the vast majority of my long-term clients made money and in many cases a lot of money, but the key to their increase has been patience. The year 2006 is a great example.

A new batch of stock-investing clients came on board during the spring of 2006. At this time, the precious metals (and base metals) were having a fantastic rally. In this real-life example I want to highlight those that were in mining stocks. There were investors who came in and immediately saw their accounts go down by mid-summer as a correction took place. Take a different client (a real client with the fictitious name Fred). Fred came on board in February 2006 with $10,000 and closed out his account 16 weeks later with $9,800. He got impatient and ended up losing $200. I feel bad that he lost $200 and I wasted my time coaching him during that brief period of time. The amazing thing is that had he continued his account, he would have been profitable in a few weeks and he would have had over a 30% gain on his positions in only a few months. A little patience could have made Fred a winner soon enough.

In a nutshell I think that impatience has been the greatest "personal" problem among investors in this decade. I'm not just saying this for long-term investors; the same goes for traders and speculators. Very often, that investment or speculative position you underwent may go down or sideways for what seems like forever. Sooner or later, if you chose wisely, others notice it too and the payoff can then be swift and impressive.

Diversification

The advice to use diversification is probably the oldest advice in investing (right after "don't loan money to your relatives" and "use a tissue . . . "). Diversification in precious metals can mean several things. It could mean spacing out your money among different metals (some precious metals and some base metals). It could mean spreading your money among different classes of investment vehicles (a mix of gold- and silver-mining stocks along with a precious metals mutual fund). For speculators it may mean deploying strategies that could benefit in up or down markets (such as using an option combination like the long straddle — more details in Chapter 15).

You can even diversify when you are speculating on a single vehicle. Last year most of my clients with commodities accounts were overwhelmingly in silver futures options. It is indeed a high-risk approach but it paid off very well. Silver that year ended up 45% and most of the silver futures options were up in triple digit percentages (sweet!). But, where possible, the options strategies involved a diversified mix of strike prices, time frames, and some *hedging,* meaning that you do something in the account that could do well if the market goes against you. Hedging is covered in greater detail in Chapters 14 and 15.

Since this book covers the world of precious metals I can cover diversification in this area, but it's important to understand that this singular area should only be a single slice of your total financial picture. You need to address other areas of your situation such as . . .

- ✔ Money in savings
- ✔ Reduce and manage liabilities (such as debt, taxes, and so on)
- ✔ Money in conventional investments such as stocks and bonds
- ✔ Real estate and other tangible assets
- ✔ Insurance and other risk-management tools and strategies
- ✔ Pension matters and retirement security

It's not a complete list but it's important to point out that whatever you do in precious metals, it doesn't happen in a vacuum. Hopefully you've looked at your entire financial picture and the economic/political landscape, and you've discovered that the best way to protect and grow your wealth in these times is by diversifying into . . . precious metals!

Diversification can be accomplished in two ways:

- ✔ Add precious metals into your portfolio to gain benefits that may not be there with other, more traditional investments.
- ✔ Have both physical metal (such as gold and/or silver bullion coins) and paper investments (such as mining stocks and precious metals ETFs), which make you diversified inside the precious metals portion of your portfolio.

Making risk your friend

I realize that after writing an entire chapter on risk that you may think that risk is a terrible concept; however, some things in this world thrive on risk. The world can be an uncertain place and many potential events and entities out there have no problem with raining some bad news on the U.S. economy in general (and your portfolio in particular). When those types of risks become evident, precious metals revert to their historical role as a safe haven. Without risk, how can you grow your money faster? It is a necessary part of your success and in many cases it is the *reason* for your success.

Risk-Management Tools

Risk is not like the weather ("Everyone talks about it but nobody does anything about it!"). It is something that you can manage and profit from. Here are some proven strategies:

- **Buy the dips.** If you bought what you think is a great stock at $10 per share and a correction hit sending it down to $8, don't just crawl under a rock and wait it out; if possible buy some more. Why not? If it is truly a great stock and your research and logic tell you it's still a solid investment, buy some more. Ultimately time will pass and the odds are good that when that stock goes to $12 or $15 or more you'll end up saying "Gee whiz! I coulda had it at $8!" No guts . . . no glory.

- **Keep cash on the sidelines.** This goes in tandem with the above point. Have some money sitting somewhere safe, liquid, and earning interest waiting for an opportunity. I tell my students that if they're ready to take the plunge with, say, $10,000 don't invest everything in one shot. Invest half now and stagger the rest in over a few weeks or a few months. Opportunities go hand-in-hand with risks so do the Boy Scout thing and be prepared.

- **Utilize stop-loss orders.** If you have a brokerage account, and it's with a major firm with a full-featured Web site, that firm has some excellent risk-management tools available for you. The most commonly used tool for keeping your portfolio's value intact is the stop-loss order. If you bought a stock at $10 then put a stop-loss order in at, say $9, or 10% below the purchase price. That way, if the stock goes up there is no limit to the upside but if the stock goes down and hits $9 a sell order is triggered and you get out. You minimize loss. A stop-loss order can be activated for a single trading day or for an extended period of time (referred to as a GTC order or good-'til-cancelled order). Stop-loss orders are a common feature in a stock brokerage account, but they may not be in a commodities brokerage account. Find out more in Chapter 16.

- **Consider put options.** Put options are a great way to protect your investment during corrections or bear markets. They can be used as insurance to protect gains or the original principal. The put option is too involved to describe at this point, but I cover it in detail in Chapter 15.

Risk (or the lack of it) isn't just where you put your money; it's also how you go about doing it. Nuff said.

Weighing Risk Against Return

After considering the risks, you may say "Gee, why have metals at all?" Well, risk is a double-edged sword. Any (and all investments) have risks, and some investments have great value because they protect you from risk, which is, perhaps, the greatest value of precious metals. Before you check out the track record, keep in mind that precious metals (especially gold and silver) are back with a vengeance in this decade. Hey . . . why do you think I'm writing this book?!

To offset some of the risk potential that is inherent in any place you put your money, use a very simple criterion, which I like to call the 10% rule: The most you should have in any single vehicle is 10% of your money. Easy! Yes . . . you could make it more precise, more customized, and more complicated. Go ahead, but it's good to have a starting point and a simple strategy before you start to tweak and "optimize."

Table 4-1 gives you a neat summary of the major ways you can get involved in precious metals based on risk. Use it to leapfrog into your investment of choice or . . . more appropriately . . . where to get more details.

Table 4-1	Precious Metals Risks and References		
Type/category	*Relative risk level*	*Most common direct type of risk*	*Chapter with details*
Bullion coins and bars (physical)	Low	Market, physical	10
Numismatic and collectible (physical)	Medium	Fraud, physical	11
Major mining companies (paper)	Low to medium	Market, political	12
Midsize mining companies (paper)	Medium to high	Market, political	12
Junior mining companies (paper)	High	Market, political	12
Mutual funds (paper)	Low	Market, political	13

(continued)

Table 4-1 *(continued)*

Type/category	Relative risk level	Most common direct type of risk	Chapter with details
Exchange-traded funds (paper)	Low/medium	Market, political	13
Futures (paper)	Highest	Market, exchange	14
Options — covered call-writing (paper)	Low	Market	15
Options — buying calls and puts (paper)	High	Market	15

Part II:
Mining the Landscape of Metals

The 5th Wave   By Rich Tennant

"Remember, I'm heavily invested in metals, so buy all the aluminum jewelry you want."

In this part . . .

So what is available to invest and speculate in? This part goes into the specific metals along with their advantages and disadvantages. You can find entire chapters devoted to specific metals such as gold, silver, platinum, uranium, and more. Get familiar with the specifics of a metal for maximum profit potential.

Chapter 5

Gold: All That Glitters

"*G*oing for the gold" and "good as gold" and many other familiar phrases mentioning gold have been around for what seems like forever but for good reason. Few things conjure up thoughts of wealth and affluence the way gold does. I don't recall any pirate in a B-movie shouting "Aargh! There's aluminum buried on that island, mateys!" On the other hand — gold . . . now we're talking. Few things have had the endurance of gold both as a metal and as a store of value. It is then the first precious metal to consider for wealth-builders today. It has through the ages become the quintessential precious metal.

Gold is an element that is found on the standard periodic table of the chemical elements. Gold is listed there with the symbol AU and the atomic number 79 (don't worry, the quiz has been cancelled). The most malleable and ductile of the metals, you could actually take a single ounce of gold and essentially stretch it out into 300 square feet (no, I don't know why you would do it but it sounds impressive). Gold is a good conductor of heat and electricity (and probably thieves as well). Because it is generally resistant to rust and corrosion gold quickly became an ideal material to fashion into jewelry, coins and therefore . . . money! The desirability of gold now became ensured.

However, as you read this book, every major society is inflating its currency at record rates. All the major currencies are subsequently losing their value slowly but surely. Because paper currencies are easily inflated, each unit of currency (dollar, euro, yen, and so on) looses value — not so for gold. As a tangible investment, use the sections in this chapter to find out how gold stacks up amidst all the investment choices available today.

The Ancient Metal of Kings

Since the early days of civilization, gold achieved its status as a symbol of wealth, a unit of exchange, and the essence of money. As an object of desire it had the qualities virtually everyone deemed as necessary to be called "money": It was a rare element that was easily carried and exchanged, and was also a universal form of money that crossed nationalities and cultures. When nations traded goods and services, gold could have easily had the slogan "Don't leave home without it." For centuries gold had a place of honor in society — even after paper currencies came on the scene.

Paper currencies (also called *fiat currencies*) started coming into usage in the days of Mesopotamia when the drachma was used. Paper currencies didn't have any intrinsic value like gold; it was more or less a *claim* to intrinsic value. In other words, a paper currency had value because it had a claim of value from another asset such as grain or precious metals. So, paper currency was really an IOU. Paper currency gained popularity because it was easier to carry and transact with versus carrying bags of coins.

Many ancient societies used paper currencies as an IOU, attaching precious metals held at a bank or secure storage facility as the asset. So, the paper currency had value as long as you could redeem the currency in precious metals such as gold. During the course of history, paper currency soon decoupled from gold and became a competitor instead. First kings and then governments saw that paper currencies were a better alternative to gold . . . or so they thought . . .

They discovered that paper currencies issued by the government had certain advantages over gold — the main advantage was that paper currencies could be printed at will while gold was not easily produced. The supply of gold throughout history only grew at about 2% per year. On the other hand, paper currencies could be produced very easily and very quickly. Although this may have been the advantage of currency, this was also currency's fatal flaw.

During the past 3,000 years, there have been literally thousands of different paper currencies that came into existence and ultimately ended up vanquished. When governments can create money by printing up endless supplies of paper currencies then the temptation to inflate becomes too great. If you can create money at will, why not? This would fill up the government's treasury early on but as more money is created, its value diminishes and it becomes worth less and less until it's . . . well, worthless. Paper currencies end up with the intrinsic value of nothing, but in the long run, gold is still gold. Gold endures. Aargh, mateys!

As the Great Depression unfolded during the 1930s, most of the major industrialized nations abandoned the gold standard in an attempt to stabilize their economies. This unfortunately opened up a long-term problem as inflating

Causing trouble with Keynes

In 1923, the economist John Maynard Keynes said that gold was "a barbarous relic." He was actually referring to the gold standard (having paper currencies backed up by gold). Considering the damage that ensued in the coming decades from his policies, I can safely call Mr. Keynes a barbarous relic right up there with Karl Marx and the pet rock.

Unfortunately, he was probably right about the gold standard because it was abandoned by governments. The abandonment of the gold standard by so many governments back then (and that no major country has one today) will probably be the primary reason why gold investors today will ultimately benefit.

their currencies became a way of life. The subsequent result is something we're all paying now . . . much higher prices.

Gold for the Record

We know that gold has a track record dating back thousands of years, but let's fast forward to recent times. In the late 1970s, gold experienced a 20-year bear market, and then the bull market started earlier in this decade. The late '70s was a time of inflation, energy problems, international conflict, and economic hardship. Hmmm . . . does that sound familiar? The conditions in the '70s definitely have a lot in common with this decade. From 2000 to mid-2007, gold has begun a bullish, upward slope. Examining gold's history and trends impacts how you decide to invest, so take a close look at the sections that follow to get an idea of gold's history as well as its promising future.

Explaining all the bull

So after a 20-year bear market, gold awakened at the start of the millennium. Gold (and other natural resources) was in the early stages of a bull market in the early part of this decade and history suggests that this gold bull market has the makings of a powerful up-trend that could easily last into the middle of the next decade (and maybe longer). Why? Several reasons explain gold's historic bull market that is unfolding in front of our eyes:

- **China and India:** These two countries alone will have an economic impact that will reverberate across the globe. Virtually every major economy and financial market will be affected, directly or indirectly. As they modernize their economies and as their collective 2.5 billion occupants increase their consumer spending, this will require much in the way of natural resources.

More specifically to the topic of gold, both of these countries have a long history of being very gold friendly. India is a huge consumer market for gold. China is loosening its formerly strict regulations and allowing its ample citizenry the ability to own gold and build wealth in a freer economy. All of this adds up to great news for gold aficionados like you and me. Whew . . . 2 billion people with money who like gold! Can you say "soaring demand"? The tears are welling up in my eyes.

✔ **The Middle East:** In spite of all the conflict and turmoil, this area has great potential. Although these countries wield economic and political clout that offers great challenges to our country, this area is yet another gold-friendly region. Many in the region are using their oil profits to buy precious metals. This means more demand.

✔ **Inflation everywhere:** Perhaps the most compelling bullish reason is that every major country in the world is growing its money supply at historically high levels; most of them are growing their money supply at double-digit rates making this an unprecedented inflationary environment. This means more dollars chasing fewer goods.

✔ **Other factors:** Other factors will play a part in gold's near and long-term future. I cover these in Chapter 19.

Telling the tale of the tape: Gold versus other investments

Gold has retained its value through good times and bad but make no mistake about it: It has an edge over other investments during difficult or uncertain economic times. In the 1930s, gold was resilient during the Great Depression; stock prices plummeted yet gold was strong. Gold did drop in price during the 1930s but not because of market supply-and-demand factors; the government banned private ownership of gold so people were essentially forced to sell it.

So you're better off comparing gold with other investments starting with the 1970s and moving onward (see Table 5-1). As of December 1974, gold ownership became legal for private citizens. Just in time . . .

Gold shines brightest during periods of inflation (especially hyper-inflation). In a period of stagflation, such as in the 1970s, hard assets (precious metals, real estate, natural resources, and so on) tend to do better than paper assets (such as stocks and bonds).

Table 5-1								Comparing Gold to Other Investments and Inflation
ASSET	**Time frame**	**Quote 1/2/00**	**Quote 6/3/07**	**Initial invest-ment**	**Ending invest-ment**	**Annual-ized rate**	**Total return for 6½**	**Did it beat Inflation?**
Gold	6½ years	282.05	650.50	$10,000	$23,063	12.90%	131.00%	Yes
Infla-tion*	6½ years	See rates	n/a		$14,755 (amount you need to retain purchas-ing power from initial invest-ment)	6.00% + (using the real rate of inflation)	45.00%	n/a
Bonds	6½ years	n/a	n/a	$10,000	$13,250	5.00%	38.00%	No
Stocks (based on DJIA)	6½ years	n/a	n/a	$10,000	$11,806	2.78%	18.00%	No
Pass-book account	6½ years	n/a	n/a	$10,000	$11,023	1.50%	9.75%	No

*Inflation isn't an asset, of course, but it's important to compare.

Table 5-1 is very telling. Certainly there are a year here and a month there that show that gold performed poorly. But then, you're not investing; you're trading or speculating. The table clearly shows that some common paper investments just couldn't keep pace with inflation. Notice the column for inflation illustrates that you would need to have at least $14,755 in 2007 just to match the same purchasing power of $10,000 in 2000. Gold resoundingly beat inflation while paper investments fell short.

By the way, the above table might make you ask about gold-mining stocks and how they fared during this period. A good barometer of the gold mining industry would be an index found at the American Stock Exchange (www.amex.com) called the Amex Gold BUGS Index (HUI). BUGS stands for "Basket of Unhedged Gold Stocks." The HUI is an index representing a batch of stocks that are gold-mining stocks that do little or no hedging. Basically, hedging is the practice of selling next year's production at this year's prices to lock in a profit. This practice is fine if gold's price is flat or declining. But if gold is rising, then the company foregoes the potential profit. Done too excessively in the wrong market conditions could spell bankruptcy for the company. I explain hedging in more detail in Chapter 12 — it's not a good practice in a rising gold bull market.

Anyway, back to the HUI. How did it perform during the time frame as compared to the other investments in the table? At the beginning of the decade, the HUI started at 73.77 (1/3/00), and it ended the day of trading on June 30, 2007, at 329.35 for a total percentage gain of 346%. In other words, a $10,000 investment in a representative portfolio of gold-mining stocks that made up the HUI index at the beginning of the decade would have been worth $44,626.27. Sweet! Yes . . . gold mining stocks are "paper" investments but fortunately they derive their value from a desirable underlying asset . . . gold.

Assuring gold's success

Gold is a tangible investment that has its own intrinsic value. I know that sounds weird, but this concept is important when you compare it to "paper" investments. All paper investments are basically "derivatives," meaning they derive their value from another entity or from a promise from another party. Think about paper investments and you see that they're essentially a promise that needs to be fulfilled. Paper investments have major risk associated with them: the risk of default by the other party. This is also referred to as *counterparty risk*. Look at the panorama of paper investments:

- ✔ **The dollar and other currencies:** When you hold a dollar in your hand, what are you holding? The dollar is a piece of paper with official printing on it. It has no intrinsic value; it is a promise of value backed up by the full faith and credit of the United States government. However, if the government (any government for that matter) prints up lots and lots of dollars (monetary inflation), the risk is that each dollar becomes diminished in value.

- ✔ **Stock:** Stock, common or otherwise, represents a piece of ownership in a public company. If the company is in business and doing well, the stock for that company will have value. But what happens if the company goes bankrupt? In that case, what will be the value of that stock beyond just being a printed piece of paper? Therefore, stock also has the risk of default.

- ✔ **Bonds:** A bond is a form of debt. If you have a bond investment, it is a piece of paper that states that another party will pay you interest and ultimately pay you back the sum of money (the principal) that was loaned. In the bond investment, you are the lender and the person obligated to pay you back is the borrower. What happens if the borrower can't (or won't) pay you back? You could lose some or all of your money in a bond. Because bonds are a paper investment they have the risk of default.

✔ **Mortgages and loans:** As this is being written in the summer of 2007, markets across the globe are in turmoil because of problems with sub-prime mortgages. A sub-prime mortgage is where money was loaned for the purpose of buying real estate (typically a home) to individuals who have below-average credit. The point is that mortgages and loans, like bonds, are paper investments that have . . . you guessed it . . . the risk of default. All situations of extending credit — whether it is a bond from a corporation, government agency or some cash you loaned to your uncle Stanley — run the risk of default.

Securing a Safe Haven from the Coming Storm

Understanding the value of gold is like understanding an umbrella without discussing rain. People quickly see the value of an umbrella when it rains. They also understand the value of buying an umbrella on a sunny day, especially when rain is forecast for the next day. To understand gold (as the umbrella) you should understand its role in bad economic conditions (the rain).

Reading and understanding inflation

As more and more people wise up to inflation, ultimately they will do something about. Although inflation has been high in recent years and has the potential to go higher, much of the public hasn't caught on. I can't say that I blame them; the responsibility for informing the public rests squarely on the shoulders of the government and the financial media. This is where I get angry. The reporting of inflation is in fact dreadful. You should understand inflation not only because of how it helps gold go up, but you should also understand it because it is a pernicious and pervasive problem. The following sections should help you sort through and gain an understanding of inflation.

Monetary inflation and price inflation

There are two types of inflation. One is a "problem" while the other is a "symptom." The "problem" is monetary inflation. That is just a fancy phrase for the government's central bank practice of printing money (actually, printing currency). The more a government prints, the more the supply of money grows.

After the money is created, it's circulated through member banks. This money makes its way into the economy through actually printed currency (such as the dollar), electronic transfer, or through loans (credit). This is the problem.

The end result of the problem of monetary inflation is the next phenomenon, price inflation. Price inflation is the one everyone talks about. That's when your uncle goes to the store and says to the sales clerk, "You're charging how much for this?!" Don't get me wrong, price inflation can be a problem for you and me but in the big picture, price inflation is the "symptom."

"Headline inflation" and the CPI

The government reports inflation as something called the CPI or Consumer Price Index. The CPI (in its various forms) is compiled and reported by the Bureau of Labor Statistics (BLS and its Web site are www.bls.gov). Unfortunately, the government keeps changing the way it's reported. An example occurred as I was writing this chapter.

I went to the BLS Web site and tried out the agency's inflation calculator. I put the year 2000 into the inflation calculator and it told me how much buying power $10,000 then would have today (August 2007). It gave me the amount of $12,096.34, which seemed too low. When I put the numbers in a financial calculator I get the annual inflation rate of 2.93%. Huh?! My dear reader, do you know any consumer necessity that has gone up only 2.93%? I didn't think so.

When you get a chance, head over to John Williams' Shadow Government Statistics (SGS) at the Web site www.ShadowStats.com. He compiles and reports inflation the way the government used to report. Starting in 1994, the federal government changed the way it reported inflation. The change lowered inflation by means such as substitution and other statistical methods that can be used to lower the inflation rate. For example, if beef is too expensive and the consumer switches to chicken, which is cheaper . . . poof! The inflation rate is lower.

Anyway, SGS reconstructs the data to report inflation in the real world. Guess what? The real rate of inflation would be a minimum of 6% (and actually higher in some years). In other words, people think that inflation is lower because of the reporting. What does all of this mean for you in this chapter on gold? It tells you that it's time to get an inflation hedge for your long-term portfolio such as gold and silver.

The core rate of inflation

In the 1990s, the Federal Reserve started reporting something called the core rate of inflation. What is it? The *core rate* is inflation excluding food and energy. Say what? You're giving me a rate of inflation that does not include food and energy? Right. Are you insane? What good is it?

Deflating inflation reporting

Why is inflation (the CPI) reported this way?

Why does the CPI under-report price inflation? First, CPI is used as a deflator for the Gross Domestic Product (GDP). The GDP is the total market value of all final goods and services produced in our country in a given year (reported for the four quarters). This number is equal to total consumer, business and government spending plus exports, less imports, and *less inflation*. The emphasis is mine. What all that means is that if inflation is lower, then the GDP is higher. Lower inflation makes the economy look better and higher inflation does not. When you calculate inflation the old way, guess what? The U.S. recently experienced a recession, and a lower inflation number makes the picture rosier.

But there is another reason that is more significant that you should know about. As you may know, millions of senior citizens get Social Security checks (and other pension checks). As you also know, they periodically get "cost-of-living" increases. Question: how do they calculate that "cost-of-living" increase? It is calculated from the official CPI. The lower the CPI, the less money government has to pay all of those retirees. The bottom line is this: If real-world inflation is running at 6% (and more) and you are a retiree getting a 2.93% increase, you are falling behind (more and more) as the months and years go by.

The reason often stated for why the Fed and the financial media report the core rate is because food and energy prices are too volatile. This is not an adequate reason since food and energy are pervasive, and they are found in every corner of a typical budget. But here is what irks me: The core rate gets more headlines than even headline inflation. The core rate gets more attention than Paris Hilton, and it's more misleading than "jumbo shrimp."

When the core rate is constantly reported, more and more people then say "Gee . . . it must be important!" and then react accordingly. For example, a headline from a major financial news Web site read: "The core rate is at a three-year low of 1.9%. The good news causes the Dow to rally 100 points." But that measly 1.9% doesn't include food and energy. The problem is that the data in recent years show most consumer necessities are up 6% to 9% and even higher. Heck . . . I'm sure that the core rate could be almost zero if you exclude more consumer necessities. While you're at it, why don't we start reporting the price of clouds? I can see a great headline: "Cumulous clouds are at 50-year low . . . Stock prices in stratospheric rise!" The bottom line is that the core rate is useless and misleading. As a financial planner I ignore it. After all, what consumer necessity has only gone up 1.9% in the past year? I rest my case.

Why should gold investors (and the public in general) be annoyed at the core rate? If investors (either by design or by accident) get the idea that inflation is only 1.9%, then they will invest accordingly. They will assume a 4% certificate of deposit or a 5% bond is helping them financially because inflation is a "lowly 1.9%." But . . . what would they do if they knew that inflation was really 6% to 9% and/or higher? Then they would adjust their portfolios to include more of those investments that would meet or beat the rate of inflation; such as . . . food and energy! In other words, consumer staples, natural resources, energy and . . . precious metals. Got gold?

Watching the dollar

The dollar is the "other side of the coin" so to speak. When people refer to "rising inflation," they're really referring to the falling value of the dollar (or other currency). When you look at the long-term chart of gold versus the U.S. dollar (Figure 5-1), you will see the dollar zig-zagging downward and gold zig-zagging upward. It stands to reason that the more and more dollars you add to the system, the less each unit is worth.

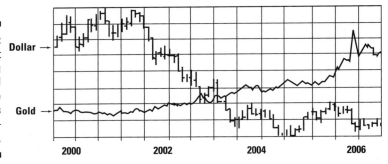

Figure 5-1: Chart juxtaposing gold and dollar price movements from 2000–July 2007.

If there was ever a picture that was worth a thousand words, it's that one. (Of course, when you adjust it for inflation, it is actually worth 1,200 words).

To monitor the price of the U.S. dollar (as a trading currency), you can go to major financial Web sites, such as Market Watch (www.marketwatch.com) and Bloomberg (www.bloomberg.com). Futures markets-related Web sites also track the dollar (along with other currencies) so you can check out places such as Future Source (www.futuresource.com), FX Street (www.fxstreet.com) and Yahoo! Finance (www.finance.yahoo.com).

Buying and owning gold

Now that you have thrown in the towel screaming, "Okay, Paul, okay! Stop it, you win! I see the light! I'll buy some gold (and aspirin)! How do I do it?" I'm glad you asked.

In the following list, I show you the many ways that you can get involved with gold:

✔ **You can buy physical gold:**

- **Bullion coins:** Gold as coins (see Chapter 10)

- **Bars and ingots:** Another form of physical gold (see Chapter 10)

- **Numismatic coins:** Gold through collectible coins (see Chapter 11)

- **Jewelry and other forms (such as collectibles):** Make sure that my wife doesn't see this (see a mention in Chapter 11). Where's that aspirin?

✔ **You can buy paper gold:**

- **Gold certificates:** More details in Chapter 10

- **Gold exchange-traded funds:** More details in Chapter 13

- **Gold mutual funds:** More details in Chapter 13

- **Gold-mining stocks:** More details in Chapter 12

- **Gold futures:** More details in Chapter 14

- **Gold futures options:** More details in Chapter 14

- **Gold indexes:** More details in Chapter 13

- **Gold managed accounts:** More details in Chapter 13

In addition to the above list is yet another way to invest in gold, and this is an interesting way to do it . . . with no downside risk. It's a gold bullion CD issued by Everbank (www.everbank.com). Called the "Marketsafe Gold Bullion CD," it's similar to regular bank certificates of deposit (it even is protected by FDIC insurance). However the investment return on it is not based on interest but is instead tied to the price of gold. If gold goes up, your CD rises in value. However, if gold goes down, the CD does not. They even have a similar CD for silver. Go to their Web site for more details.

The Gold Market

Gold is a worldwide market and the long-term supply-and-demand fundamentals are very bullish. The following sections give you some insights into the gold market.

Gold market data and information

The more you know about what is happening in the gold market, the better prepared you will be to profit. The following organizations provide extensive research and data on the yellow metal:

- ✔ **Gold Fields Mineral Services (GFMS):** Gold Fields Mineral Services is probably the leading global consulting firm doing specialized research on the world's precious metals market, specifically gold, silver, platinum, and palladium. GFMS is based in the United Kingdom and has offices and representatives stretched across the globe.

  The lead executives, Philip Klapwijk (Chairman) and Paul Walker (CEO), are active on the conference circuit and their articles are regularly published and worth reading. Find out more about GFMS at their Web site (www.gfms.co.uk).

- ✔ **CPM Group:** CPM is another leading consulting firm that specializes in commodities and precious metals research. They provide extensive research on the precious metals markets and publish their work in several annual yearbooks: The Gold Yearbook, The Silver Yearbook, and The Platinum Group Metals Yearbook. The CPM Group was founded in 1986 by Jeffrey Christian, a metals analyst and the author of the book *Commodities Rising*. Its Web site is www.cpmgroup.com.

- ✔ **Gold Anti-Trust Action Committee (GATA):** The Gold Anti-Trust Action Committee was formed in January 1999 by well-known gold analyst Bill Murphy. GATA advocates and undertakes litigation against illegal collusion between financial institutions (such as central banks and large brokerage firms) that intervene in the gold market. GATA's aim is to foster a more free market for gold since the central banks and brokerage firms have on an ongoing basis managed to control the price and supply of gold and related financial securities. More information on the topic of political market intervention in precious metals is in Chapter 19. GATA's Web site is www.gata.org.

- ✔ **World Gold Council (WGC):** The World Gold Council was created by gold mining companies and refiners in 1987. Its mission is the promotion of investment and usage of gold by consumers, investors, industry, and central banks. The WGC's Web site is www.gold.org.

- ✔ **National Mining Association (NMA):** The NMA is the voice of the American mining industry in Washington, D.C. and the only national trade organization that represents the interests of mining before Congress, the Administration, federal agencies, the judiciary and the media (www.nma.org).

Industrial supply and demand

The major determinants for gold's performance (its price going up or down) boil down to inflation and supply-and-demand factors.

Total demand and supply

Let me give you the big picture first. In recent years, total average annual worldwide demand has been in the range of 3,800 to 4,000 tons. Total average annual worldwide supply (coming on board from mines, and so on) is in the range of 2,500 to 2,700 tons. Because of the slight reporting differences from several sources, I thought it best to give a reliable range. The bottom line is that there is an annual shortfall in the range of 1,100 to 1,500 tons, depending on which numbers you calculate.

No matter which way you slice it, the big picture in the long-term is that supply-and-demand factors for gold are bullish.

Demand

Demand for gold comes from two places: investment and industry. Industrial demand is roughly 10% to 15% of worldwide annual demand. Gold has unique properties which mean more applications of it in technology, healthcare, and other vital industries. Gold is extremely resistant to corrosion and it has high thermal and electrical conductivity. Therefore, it is an excellent component in electrical devices. Gold is also used in medical equipment because it is resistant to bacteria. New uses for gold have been found in pollution control equipment and fuel cells. In addition, gold has promising application in the new area of nanotechnology. The practical uses for gold keep growing. (Hmmm . . . I wonder how much they would offer me for my wife's jewelry?)

For current news on the many new ways gold is being used in industry and medicine, check out gold's growing uses UtiliseGold at `www.utilise gold.com`.

Supply

Gold is mined on all the major continents. Specifically, it has been most plentiful in places such as South Africa, Canada, the United States, Russia, and Australia. There is no mining in Antarctica but then again, why would you want to? At present there are over 400 mines active across the globe. With supply from these mines coming on board in the range of 2,500-2,700 tons, this means that the world supply of gold increases by roughly 2% per year. That's significant. Why?

For thousands of years, gold was (and is) considered money. It has outlived thousands of currencies, and in due course, it will outlive more currencies. Part of the reason for its durability as a desirable form of money is directly tied to its rarity. You can create lots of paper currencies almost instantly but you can only increase gold by a mere 2% per year. It is easy to inflate man-made currencies and history tells you this has happened very, very often. It is happening today! But . . . it's not easy to inflate gold.

Although new mines do come on board every year, it doesn't have a significant impact on total worldwide supply since old mines do close down after no more gold is found. In addition, it often takes mines 5 to 10 years (sometimes longer) to go from discovery to production. Although increased market prices for gold are indeed an incentive for more gold exploration, it doesn't happen quickly. It does take years to translate higher market prices into new gold mine discovery and production.

When the price of gold does go up significantly, gold "scrap" does find its way into the marketplace. Scrap is basically recycled gold. As the price of gold goes up, it becomes more economical to recover scrap to refine it and reuse it.

Investment demand

Since the beginning of the decade and obviously due to gold's bull market, investment demand has grown tremendously. Actually, according to market studies by the CPM Group, a very unusual situation had arisen during 2006–07. In that time, private investors (as a collective group) owned more gold than the group of central bankers. The first time in world history!

Much of the new investment demand comes from two sources: the popularity of precious metals-related exchange-traded funds (ETFs: more on those in Chapter 13) and from international investors such as from China, India, and the Middle East. More and more investors are seeing the big picture, and they understand that gold has unique advantages and is a good part of a well-balanced portfolio.

Central banks

Central banks play a pivotal role in the gold market and sometimes that role can be a controversial one. It stems from the "love/hate relationship" that central banks have with gold (maybe they need therapy). On the one hand, central banks are the single largest holders of gold in the world. Most governments have gold as an asset in their official reserves. Yet, on the other hand, gold is a "competitor" to the currency that is issued by the central bank on behalf of its government.

You'll notice that whenever gold is rallying, one or more of the central banks will either announce a large sale of gold or actually sell a large amount of it. And they have a considerable amount of the yellow metal; it is estimated that central banks along with other governmental institutions (such as the International Monetary Fund) hold roughly 20% of the world's above-ground supplies of gold. This stockpile is approximately 30,000 tons (that's tons, not ounces) of gold.

Keep in mind that not all central banks work in unison. Some are buying gold while others are selling. Although the central banks have much autonomy about how to manage their individual gold reserves, most of the major central banks have agreed to do their buying and selling according to guidelines stipulated in agreements such as the 1999 Central Bank Gold Agreement (CBGA). You do see some interesting developments in recent years.

Central banks in "the West" such as the U.S. Federal Reserve and central banks in Western European governments (such as the United Kingdom, France, Spain, and others) have been net sellers of gold. Meanwhile, central banks in the East (both middle and far) have been net buyers of gold. This tends to generally reflect attitudes about gold in both spheres. Folks in Asia and peripheral regions as groups embrace gold far more than those in North America and Europe. As a total group, it has been a net seller which was a contributing factor to some of those steep corrections that have been experienced by precious metals investors in recent years.

Central banks don't have to just buy and sell their gold to influence the market. They are also involved in activities such as gold derivatives and in leasing gold. More about some of these activities in Chapter 19.

Central banks report their gold position and activities to the International Monetary Fund (IMF) and you view some of this information at the Web site at `www.imf.org`.

Gold Bugs

The term "gold bugs" has sometimes been used as a pejorative but many veteran gold analysts and well-known commentators have come to wear the name like a badge of honor. The following is really like a "hall of fame" for the world of gold since some of these stalwarts (actually most of them) were successful and active back in the late 1970s.

What these folks have in common goes beyond an affinity for gold investing. Through the years they have offered research and insights that benefited the gold-investing community for decades. Some of their forecasts became legendary. Yes, they have had their misses, but in the investment world, when you have far more hits than misses and keep doing it for years and even decades, this is something we can't ignore.

To me, another reason for checking them out is that I like to know that you just don't talk the talk you need to also walk the walk. If I know that the person involved made a lot of money on his or her own advice, that goes much further than people who tell you to invest in X and then you find out that they made most of their money by selling reports and newsletters about "How to invest in X." In other words, these folks are primarily self-made in their financial independence due to the adherence and application of their own knowledge.

Drum roll please! Here is a pantheon of gold bugs worthy of the name:

- Howard Ruff (www.rufftimes.com)
- Jay Taylor (www.miningstocks.com)
- Doug Casey (www.kitcocasey.com)
- James Sinclair (www.jsmineset.com)
- James Dines (www.dinesletter.com)
- James Turk (www.goldmoney.com)
- Mary Anne and Pamela Aden (www.adenforecast.com)
- Harry Schultz (www.hsletter.com)
- Adam Hamilton (www.zealllc.com)

By the way, they're still all more bullish than ever on gold (and other precious metals). 'Nuff said.

Gold Investing Resources

The gold market can be a fascinating one and if your money is riding on it then it better be a well-researched one as well. Here are some solid Web sites for gold investors:

- Le Metro Pole Café (www.lemetropolecafe.com)
- Gold Sheet Mining Directory (www.goldsheetlinks.com)
- Gold Eagle (www.gold-eagle.com)
- About.com (www.metals.about.com)

- The Gold Institute (www.goldinstitute.org)
- 3 2 1 Gold (www.321gold.com)
- World Gold Council — Investor information (www.invest.gold.org)

Staying informed about the gold price and getting the latest news:

- Kitco (www.kitco.com)
- The Bullion Desk (www.thebulliondesk.com)
- INO.com (www.ino.com)

Pssst. I got another tip for you. If gold goes as high as I think it will, you can actually celebrate with . . . uh . . . an adult beverage that has actual gold flakes in it! Of course, if gold doesn't do well . . . uh . . . you still have the adult beverage.

Chapter 6

Discovering the Secret of Silver

In This Chapter

▶ Understanding a unique metal with dual benefits

▶ Finding silver market information

▶ Exploring a future with a silver lining

Silver is hands-down my favorite precious metal. Maybe it's because it's more affordable than gold. Maybe it's because I like the color better. Oh wait . . . now I remember: When I buy my wife silver jewelry instead of gold I still have some cash left to buy something for my boys! Well, really . . . I like silver most because of the great profit potential that I think it will generate. It has served me well in recent years and there is no reason why that shouldn't continue (and get better!) in the coming years.

Silver started the decade at around $4 and in early 2007 silver hit $14 an ounce — it did very well. I believe that the conditions for it to beat its old high ($50 in 1980) are excellent in the next few years. I think devoting a whole chapter to the poor man's gold is a good idea. Silver's future profit potential is truly its silver lining (you were expecting that line . . . right?).

Understanding the Hybrid Potentials of Silver

Silver is a fascinating metal and it has the unique, dual quality of being both a monetary metal (used as money) and an industrial metal. This lies at the heart of its potential. There are countries that are reconsidering the use of silver again in their currencies. Meanwhile, more uses for silver in technology and healthcare mean more opportunities for silver investors and speculators.

Silver is unique in that no other metal combines strength with a softness that allows it to be formed and stretched. You would be hard-pressed to find a metal that conducts electricity as well or is as pliable, corrosion, or fatigue resistant. Nothing else has such high-tensile strength, is wear resistant, has such a long functional life, or is as light sensitive. Silver endures extreme temperatures, conducts heat, reflects light, provides catalytic action, and has a great reputation for its bactericidal qualities. It alloys (easily combines with other metals) and helps to reduce friction. Due to its exceptional properties and reasonable price, there is no substitute for silver.

Monetary uses for silver

For centuries gold and silver were used as money. Silver was probably more widely used since gold was too expensive for day-to-day use. Silver was not as valuable as gold as a monetary unit so it served nicely for the smaller transactions. Silver was used in ancient times and it was part of our currency in the United States since colonial times. It was regularly used in our coinage (up to 90% silver content) as dimes, quarters, and higher denominations until 1964. From 1965 onward, dimes and quarters ceased having silver content. Halves commemorating John F. Kennedy contained 40% silver content from 1965 to 1970 before joining the other coins and only having base metal (such as copper and nickel) in their composition. If you see any of those old silver-content coins appear in your change — keep them! A silver coin is far more valuable for its metal content than for its face value. Silver (and gold) coins are covered in Chapter 11.

Silver is rarely used in coinage for general circulation today. The exception was for special collectible issues that the U.S. Mint did from time to time. In 1986 the U.S. Mint did start minting the silver eagle, which quickly became prized by collectors and investors (see Chapter 10).

Industrial uses for silver

In the past, the primary engine of industrial usage for silver was photography. As traditional photography started slowly going the way of the buggy and the corset, the true strength and versatility of silver started to come forward. David Morgan of www.silver-investor.com makes a fascinating point about silver: There are more new patents with silver than with all other metals combined. There are thousands of industrial applications for silver. As a matter of fact, out of all the commodities in use in our economy, only petroleum is used in more different ways. (Of course, I am also bullish on energy.)

The largest use of silver comes from industrial demand, jewelry/silverware, and photography, in that order. Although photography has gotten all the attention in recent years, of the three major categories, it is the smallest.

Sorting through silver's many uses

The multiple uses for silver started growing almost exponentially as industry figured out that silver's unique properties made it an ideal component in a broad array of products. Here is a partial list:

- Bactericide
- Batteries
- Bearings
- Brazing and soldering
- Catalysts
- Clothing (lining to kill bacteria)
- Coins
- Computer components
- Electrical
- Electronics
- Electroplating
- Jewelry and silverware
- Medical applications
- Mirrors and coatings
- Nanotechnology
- Photography
- Plasma screens
- Solar energy
- Super conductivity
- Surgical instruments
- Washing machines
- Water purification

Yes . . . it is a partial list and it keeps growing.

Industrial demand for silver makes up 43% of total demand, and this area is also the fastest *growing* area of silver demand. The industrial portion of the market is growing at about 2% per year. It is important to understand that in almost all instances, the amount of silver used in a cellphone, laptop computer, or microwave oven is so small that it cannot be recovered. For all practical purposes, the silver used in these applications is lost and unrecoverable.

This type of demand is called *price-inelastic* by economists. The small amount of silver that is used makes it an insignificant factor in the price of the product. The amount of silver used in the manufacture of a battery, an automobile, a computer, or cellphone is insignificant when compared to the price of labor and other materials. A doubling in the price of silver would not affect, for example, what Honda uses in making an automobile. Since the price of silver has such a small relationship to the cost of the finished product, there is really no substitute. If the price of silver went to over $100 per ounce, for example, the only possible substitute for silver would be palladium or platinum, both costing much more than $100 per ounce.

Researching Silver

Making fantastic profits in anything is not easy, and silver is no different. Even when you think a speculation is a slam dunk, it can backfire in the short term. You need to be armed with market intelligence. You need the facts, and if you're seeking options, then get informed opinions from long-established sources.

Sources of data

Sometimes opinions are no more valuable than fiat currency (normally a solid laugh line at precious metals conferences). In other words, get the facts. Silver investors and speculators fortunately live in the best of times when it comes to gaining access to reliable information. With some time and effort at a good business library and the Internet, of course, you can find the information necessary to render an informed decision. The following sections list the top sources.

The Silver Institute

Established in 1971, the Washington, D.C.–based Silver Institute is a nonprofit international association with members that span across the silver industry. They regularly track the sources and uses of silver and they publish their research in their World Silver Survey. Their recent research was made available in reports covering topics such as the Chinese Silver market and a study of stockpiles of silver around the world. More information can be found at www.silverinstitute.org.

The CPM Group

The CPM Group is a research and consulting firm that specializes in the metals industry. They produced annual studies in the individual precious metals (called yearbooks) and the most relevant one is the CPM Silver Year Book 2007. Their research includes data on production, consumption, and silver output. More details are at www.cpmgroup.com.

New York Mercantile Exchange

For those seeking market data on silver futures and options, go to the source: the New York Mercantile Exchange (NYMEX) at www.nymex.com. Silver is just one commodity traded at NYMEX but you can find relevant information on silver and this fast-moving market.

SilverStrategies.com

For those researching silver companies, market data, and relevant news, Silver analyst Sean Rhakimov's www.silverstrategies.com is worth a visit.

Sources of informed opinion

As silver and precious metals in general become more popular and visible, more and more newsletters and advisory services start to come out of the woodwork. When a new batch of "experts" and "gurus" emerge, that might tell you that it's a popular market, but it becomes a dart-throwing exercise when it comes to choosing the ones that have helpful guidance and information. Who do you turn to?

I think it's good to turn to those sources that have had a consistent track record and have followed the silver market for years in both up and down markets. Two that specialize in silver are the Morgan Report (at silver analyst David Morgan's www.silver-investor.com) and the Silver Stock Report (silver analyst Jason Hommel at www.silverstockreport.com). Ted Butler's essays and features are provided at www.investmentrarities.com. Doug Casey's International Speculator also does a great job covering silver stocks (www.kitcocasey.com).

There are many gold advisory services that also report on silver opportunities and they can be found in Chapter 5.

Owning Silver

To maximize your profit opportunities and provide some safety and a measure of diversification, get involved in both physical ownership as well as paper investments of silver. Owning physical silver is a solid, long-term, buy-'n'-hold strategy that should work well. For those seeking more growth (along with more risk) then consider paper investments such as mining stocks. The following sections give you some pointers for both.

Physical silver

Silver experts Ted Butler and Jason Hommel are staunch proponents of buying and holding physical silver. They enumerate the benefits in their respective Web sites. The easiest way to get into silver is through bullion silver coins such as the U.S. silver eagle. It is produced by the U.S. Mint (www.usmint.gov), and it's readily available either directly or through coin shops and dealers.

The benefit of buying bullion is that the price you pay primarily covers the silver content. A silver eagle, as you shop around, could be bought for whatever is the spot price of silver plus a premium of 5% to 12% to cover production and marketing costs. Therefore, if the spot price of silver is $10, you could find the cost for a silver eagle to vary from $10.50 to $11.20. Be careful as there are dealers that have higher premiums, so do shop around.

There are silver commemoratives, medallions, ingots, and other collectibles available, but you'll find that the price of what you are buying will cost far more than the silver content. You may not even get as much silver as you think. The standard is .999 fine silver so find out as quality can vary.

Another area is numismatic coins but that is a different animal altogether. Numismatics (or collectible coins) are covered in Chapter 11.

Keep in mind that another benefit of buying bullion coins such as silver eagles is that you can easily sell them as it's a large and liquid market. Collectibles and numismatic coins require more diligence, effort, and expertise in both buying and selling them.

Paper silver

You can invest in silver without actually holding or storing the physical silver. Many people buy silver through things such as silver certificates or through brokers and dealers that store it for you and provide proof of ownership such as through a warehouse receipt. There are some pitfalls so you would need to choose reputable dealers that document the existence of the silver it holds on your behalf. It was a common problem where people bought silver (or so they thought) only to find out that they paid good money for a piece of paper that was not backed up by actual physical silver allegedly held by the vendor.

Another way to invest in silver is through buying common stock of silver-mining companies. There are a few large silver-mining companies and several dozen smaller companies. To invest in them it requires not only some due diligence in understanding the silver market but also about public companies in general. You need to know how they make money and what their financials tell you. In addition you need to know what the pitfalls are. Mining stocks are covered in Chapter 12, and there are companies that can be good choices for either conservative or aggressive investors.

A safer way to invest in mining stocks (especially for beginning or novice investors) is through mutual funds and exchange-traded funds (ETFs), and Chapter 13 gives you more guidance. You will find very few (if any) mutual funds that are purely into silver-mining stocks, but there are many precious metals funds that have silver mining stocks as part of their portfolio.

Lastly, I can mention the most aggressive and/or speculative way to be in paper silver vehicles and that would be futures and options. This requires much more due diligence and a greater tolerance to risk. Fortunately, books like this will help (you're welcome!). Futures are covered in Chapter 14 and options are covered in Chapter 15. Between the two, my personal preference for speculating is options since I can still get "the bang for my buck" similar to futures but I can limit my risk.

Silver's Compelling Future

According to the CPM group's 2003 silver survey, there are approximately 400 million ounces of silver bullion and 2 billion ounces of gold bullion available. Before moving on, it is important to qualify this fact. First, this comparison is between gold bullion and silver bullion. In both cases, I am not talking about jewelry or art forms of the metals. However, to clarify the point, if silver coinage was added to the silver bullion, the total would still be approximately one billion ounces. This is less than one-third of the gold supply, if we count both gold coin and gold bullion. Since then, the supply of gold has increased and the supply of silver has decreased. The following sections give you an idea of what factors into silver's compelling future.

Market fundamentals

There was a time when silver was plentiful. At one time the U.S. government had over two billion ounces of silver stockpiled during the 20th century, but as of 2003 it has no more. As we consume more silver than we produce, a major drawdown of supplies begins to occur. You begin to see how rare silver is when you compare it to gold. Virtually all the gold that has been mined is still around in some form. You can find large supplies held by central banks, and gold ownership (through coins and jewelry) is at an all-time high. Silver, though, is a different story.

The demand side

One of the most incredible things regarding silver is that demand for it has outstripped supply for 15 straight years. This trend is projected to continue for at least the next several years. Annual silver supply deficits have run as high as 200 million ounces in boom years, and as low as 70 million ounces in years of recession. It is important to realize that even in years of decreased silver demand, the mining supply on an annual basis did not meet demand. There is nothing more bullish for a commodity than such a deficit condition, a condition where demand is greater than available supply.

The supply side

Because the vast majority of silver is mined as a byproduct of mining lead, zinc, copper and gold, the price of silver can be sensitive to the amount produced as a byproduct. This, however, is a condition that has a flipside. Since the total amount of silver produced by mining has fallen short of total demand and the price of silver has started to climb, there is little incentive for base metal miners to rev up their production because the silver produced is not instrumental to their business. The ongoing commodity boom will continue, and base metal production will increase in the future, but it is highly doubtful that the increase will keep pace with the increasing demand for silver for many years.

The world needs more silver

All too often, we think in terms of America and what we produce and consume. In today's economy, we're no longer the 800-pound gorilla. While America is slowing down, the rest of the world is speeding up. Precious metals (and commodities in general) could easily have a long-term bull market with just two reasons: China and India (and the rest of Asia).

Asia will be a tremendous factor for two reasons. First — industrial demand — Asia's growth and accelerating industrialization mean it will need more natural resources. Remember that precious metals are finite natural resources. Secondly, China and India have always embraced gold and silver in their cultural and social circles. There is natural demand that will only increase as their respective nations become more affluent. More details about the international demand for precious metals are in Chapter 18.

The Legends of Silver

Every market has those trailblazers that stay on the cutting edge of market movements. The stock market had Benjamin Graham and Warren Buffet. The mutual fund industry had Peter Lynch and John Bogle. If you had followed their research, you would have very done well in those markets. The precious metals market is no different. The following sections profile the trailblazers in silver.

Jerome Smith

One of the first legendary "silver bulls" was Jerome Smith. In 1972 his book *Silver Profits in the Seventies* accurately detailed the bull market for silver during the high-flying '70s. At the time, he was also very accurate about the rise in gold and other precious metals. His writings influenced the Hunt brothers in their infamous attempt at cornering the silver market.

Smith expected silver to explode during the 1980s but was way off the mark. The reasons he cited are probably more appropriate now, a quarter-century later, since he did not foresee how much silver would enter the marketplace from government sources that depressed the price. He also didn't foresee the extent that large financial firms in the silver market would be active in forcing the price of silver down (referred to as *shorting* the market). Fortunately, his work influenced others to continue tracking the silver market and its great potential today.

Ted Butler and the silver shortage

In recent years, landmark studies and research into the silver market was undertaken by silver analyst Ted Butler (www.butlerresearch.com). He has been one of the very few analysts to track and research the silver market on a full-time basis. Although he usually doesn't issue specific forecasts and recommendations, he was correct on a consistent basis on warnings about problems with silver and other metals. He warned about practices such as forward selling, naked short selling, and metals leasing. In recent years, he has warned investors and the Commodities Futures Trading Commission (CFTC) about a huge and untenable short position in silver at the New York Mercantile Exchange (NYMEX).

Butler is on record that silver has the long-term realistic potential of going into three figures ($100 or more). He also embraces Jerome Smith's forecasts that silver has an outside chance of becoming more valuable than gold when the world marketplace realizes that silver is actually more rare than gold. Ted Butler cites the following points as the fundamental reasons to be bullish on silver long term:

- ✔ **The huge supply and demand imbalance:** As new industrial applications need silver and as the rest of the world continues modernizing and growing (again, places like China and India), demand will outpace (and is already outpacing) new silver being mined. Long-term, a very bullish condition.

- ✔ **The huge short position in silver:** In the futures market, you can go long in silver (buying a contract) or you can go short (selling a contract) as a market strategy. Going short in silver means that you borrow silver (in this case a futures contract) and sell it immediately, and if the price of silver goes down, you can buy back the silver at a profit. The key point to remember is that going short involves a liability: If you borrow, then you must return or pay back. If you go short, say, 100 silver futures contracts, then ultimately you need to buy back those 100 contracts to complete your transaction (covering your short position). Don't worry if you don't get it immediately — I explain it further in Chapter 14. The bottom line is that Butler warns that there is a huge short position in silver that exceeds the known inventory of silver available — a very unusual situation that could prove to be explosive for the price of silver in due course. In early May 2007, records showed a short position at NYMEX of over 508 million ounces of silver while total silver on hand at NYMEX-approved storage facilities was about 132 million ounces. In other words, there's not enough silver to cover potential purchases of silver. When a forced purchase in a shorting situation occurs, this is called a *short squeeze*. Long term, this is a very bullish condition.

✔ **Unbacked silver bank certificates:** Over a period of decades, approximately one billion silver bank certificates were sold with the provision that they can be converted to physical silver. However, there aren't a billion ounces available to back up these certificates. In due course, as some certificates begin to be converted, what happens when the point is reached and no silver is readily available? The signal to the marketplace is that there is a silver shortage and . . . yes . . . this would be yet another reason that silver would increase in price.

The above major points come from Ted Butler's research and from publicly available data. Butler makes additional points about the bullish case for silver but you get the picture.

David Morgan

For years David Morgan of www.silver-investor.com tracked the silver market and offered cutting-edge research on silver and especially silver mining companies. His book *Get the Skinny on Silver* (www.gettheskinnyon silver.com) quickly became a key reference source especially for beginners. Although Morgan made a number of accurate short-term forecasts for silver, his research indicates that long-term silver is heading for an all-time high price since the supply-and-demand fundamentals are so strong. After surveying the industry, Morgan notes that silver supplies are at historically low levels to the point that if any large investor or industrial firm wanted a significant supply of silver that it could trigger a buying panic causing the price of silver to spike (move upward very quickly). Figure 6-1 illustrates the latest supply and demand data for silver.

	1996	1997	1998	1999	2000	2001	2002	2003	2004	2005
Supply										
Mine Production	490.9	520.0	542.0	556.7	590.7	606.4	596.5	601.0	620.4	641.6
Net Government Sales	18.9	-	33.5	97.2	60.3	63.0	59.2	90.6	66.5	68.0
Old Silver Scrap	158.4	169.3	193.9	181.2	180.4	182.4	187.0	183.6	181.2	187.3
Producer Hedging	-	68.1	65.5	-	-	18.9	-	-	10.0	15.1
Implied Net Disinvestment	142.8	86.2	53.1	48.8	100.0	-	20.6	-	-	-
Total Supply	811.1	843.6	829.1	883.9	931.3	870.7	863.3	875.2	878.1	911.9
Demand										
Fabrication										
Industrial Applications	297.7	320.8	316.4	339.2	375.5	336.4	340.2	350.8	368.3	409.3
Photography	210.1	217.4	225.4	227.9	218.3	213.1	204.3	192.9	181.0	164.8
Jewelry & Silverware	263.7	274.4	259.4	271.7	278.1	286.9	262.4	273.8	247.8	249.6
Coins & Metals	25.2	30.4	27.8	29.1	32.1	30.5	31.6	35.6	42.3	40.6
Total Fabrication	796.8	842.9	829.1	867.8	903.9	866.8	839.5	853.0	839.4	864.4
Net Government Purchases	-	0.7	-	-	-	-	-	-	-	-
Producer De-Hedging	14.3	-	-	16.0	27.4	-	24.8	20.9	-	-
Implied Net Investment	-	-	-	-	-	3.9	-	1.3	38.7	47.5
Total Demand	811.1	843.6	829.1	883.9	931.3	870.7	863.3	875.2	878.1	911.9
Silver Price (London US$/oz)	5.199	4.897	5.544	5.220	4.951	4.370	4.599	4.879	6.658	7.312

Figure 6-1:
Silver
supply and
demand.

Chapter 7

Platinum and Palladium

In This Chapter

▶ What the platinum group metals are

▶ The ways you can invest and speculate in them

▶ Resources to help you choose

*P*latinum and palladium (and four other metals) are part of what is called the platinum metals group. Before you think it's a new rock group, think again (well, now that I think about, it really is a rock group). The six metals that are in this group are elements that are found together in the table of elements and as metals have much in common. I'll get to the six metals later in this chapter. Because platinum is the most significant metal of the group, it became the nominal leader of the pack. Platinum and palladium are the most common metals in this group for investors, traders, and speculators.

Naturally occurring platinum and platinum-rich alloys have been known for a long time. Centuries ago, Spanish explorers came across the metal and named it *platina* or "little silver" when they first discovered it in South America. They initially didn't want platinum because they regarded it as an impurity in the more desirable silver they were mining. Oh, man . . . if they only knew! As I write this chapter, I just checked the price; platinum (at $1,270 an ounce) is about 100 times pricier than silver. I could certainly tolerate that kind of four-figure impurity in my portfolio.

In this chapter, I explore how this rock group can help you diversify your portfolio, much like other precious metals, such as gold and silver.

The Platinum Group Metals

The platinum group metals include platinum, palladium, rhodium, osmium, ruthenium, and iridium. We need 'em, we use 'em, so we might as well invest in 'em. Okay, so let's see 'em. Drum roll please . . .

Platinum

Platinum is the principal and most well-known metal of the group. It is generally used in jewelry and in industry. The largest single demand for platinum is in jewelry, which accounts for 51% of the demand. The second largest use is in the automotive field (29%) for components such as catalytic converters and related parts. Lastly, the remaining demand is used in a small variety of other uses. Among them: electronics, chemical and petroleum refining catalysts, and collectibles such as bullion coins.

Platinum is an excellent conductor of electricity and does not corrode. It is among the world's scarcest metals where new mine production amounts to approximately 5 million (troy) ounces a year. To put that relatively small figure into perspective, gold production is in the general neighborhood of 80-85 million ounces per year. Platinum is mainly found in places like South Africa, Russia and North America. In fact, South Africa accounts for about 80% of the world's supply.

Because of platinum's limited supply and the world's continuous demand for it, investment interest should keep growing as well. In case you were wondering, the platinum credit cards that have been so popular lately don't have any platinum.

Palladium

Number two on the list is palladium but it keeps trying harder. Palladium has become more established in recent years as a useful industrial material as well as a very tradeable metal on the world market. In industry, it is used in diverse areas such as automotive, electronics, oil refining, dentistry, chemical, medicine, fuel cells, hydrogen purification, water treatment, and more. Consumers are buying more palladium jewelry and coinage.

And you thought that it was just a name for a popular dance club!

Rhodium

Rhodium was discovered by the English chemist William Hyde Wollaston in 1803. That was a hot year for him since he also discovered palladium (alas, since he didn't discover the "futures market," he couldn't discover some real money either). Rhodium tends to occur along with deposits of platinum and is primarily obtained as a byproduct of mining and refining platinum. (Rhodium was also discovered as a byproduct in nickel in Canada.) The leading source of rhodium is South Africa with 60% of the known supply (Russia is second). It is rare; only about 7 to 8 tons hit the market every year.

Rhodium is used to make electrical contacts, as jewelry, and in catalytic converters, but is most frequently used in tandem (as an alloy) with other materials like . . . you guessed it . . . platinum and palladium. These alloys, in turn, are used in industry such as aerospace and manufacturing.

For me, the most eye-popping point about rhodium is its price tag: over $6,000 an ounce as of September 2007. (It could have been had for only $380 an ounce in 2002.) When she asks, I'll tell my wife to put a little rhodium in my Christmas stocking.

Osmium

Osmium. Yep. It's real stuff. I'm trying hard to write something clever about it but to no avail. Anyway, it's a lustrous, bluish white metal that is extremely hard and it has the highest melting point and the lowest vapor pressure of the platinum group (tell everyone at your next cocktail party).

Osmium is used to produce very hard alloys with other metals of the platinum group so that it can be used in a variety of products ranging from fountain pen tips to pacemakers and medical instrumentation replacement valves.

Iridium

Ya see, when you do a book like this, you have to occasionally scrape the bottom of the barrel. You get to write (if only briefly) about things that sound like they came out of an old episode of Star Trek (Captain Kirrrrk! We got no powerrrr. We need more iridium crystals . . . and a better script!)

Iridium is found with platinum and other platinum group metals in alluvial deposits. Iridium is the most corrosion-resistant metal known. It is used in medical instrumentation, electrical contacts, and as a hardening agent for other metals.

Ruthenium

Ruthenium is also found with the other platinum group metals found in locations such as the Ural Mountains and in North and South America. It is used as a hardening agent for platinum and palladium. When it is alloyed with these metals it is then used to make electrical contacts for severe wear resistance. Ruthenium, when added to titanium, makes its corrosion resistance 100 times better. You go, Ruth!

I've been told that when you combine osmium, iridium and ruthenium as an alloy to cement you get . . . stadium. I could be wrong . . .

The investor's best choices in the group

Let's face it; at this point in time it is not practical for folks to get involved in osmium, iridium, ruthenium, and the expensive kid on the block, rhodium. There are probably some publicly traded companies that deal with them but the real action and investment activity are with the Lennon and McCartney of the group, platinum and palladium. These two have much profit potential to offer investors as well as more different vehicles.

Platinum's performance

This metal has lots of style and grace but more importantly, it packs a nice investment punch. Figure 7-1 gives you a snapshot of what it has done recently . . .

Figure 7-1:
Five-year
chart of
platinum on
an upward
slope.

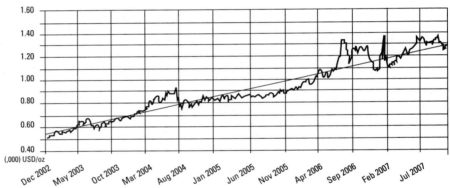

Platinum may be attractive as jewelry but it's downright gorgeous as an investor's price chart. As you can see, platinum had a nice, steady up-move during the five years ending September 2007. The metal was under $500 during 2002 and it hit $1,270 for a total percentage gain of well over 150%. During the spring 2006 rally, it did trade as high as $1,355.

As investment vehicles, platinum (as well as palladium) has the same dual qualities (investment demand and industrial demand) as silver. However, the composition is a little different.

It can be argued that silver's investment and industrial demands are roughly equal while platinum and palladium have much more industrial demand than investment demand. Therefore, platinum and palladium (and the other metals in this group) are much more correlated with economic growth. In this respect, the platinum group metals more closely approximate the movement of base metals.

In the "big picture," however, the platinum group metals should continue to flourish in the current investment climate. As supply and demand factors and inflationary pressures continue, investment and industrial demand will continue to enhance potential returns for platinum group metals investors and speculators.

Palladium's performance

Of all the major tradeable and investable metals, palladium had the greatest swings in the course of its past five years of market action. Just look at the chart in Figure 7-2.

Figure 7-2: Five-year chart of palladium.

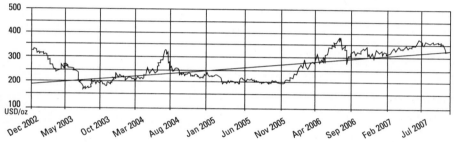

As the chart shows, palladium has been a great trading vehicle since there was such a roller-coaster ride for its price. For those who don't mind the short-term volatility, palladium would have been an ideal vehicle for those who like to employ option strategies such as a straddle. A straddle is basically two bets that you make simultaneously on the same asset: one bet that it is going down and the second that it is going up. This is done by buying both a call option and a put option at the same time on the same asset (see Chapter 15 for more details on the straddle). When palladium shot up, you could have cashed in the call at a nice profit. Then later on as palladium plummeted, you could have made another profit by cashing out the put option.

For long-term investors, you would have thought that nothing happened with the metal at all. Palladium in September 2002 was in the mid $300s range, and by September 2007 it still ended up in the mid $300s range. Oh well. You can't always have a runaway success. Fortunately, the mining stocks involved in platinum and palladium did manage to beat the stock market averages.

Platinum Group Investment Vehicles

Moving forward, the next issue is if you are going to get involved with platinum and palladium, how can you proceed? The following sections discuss the several ways you can get involved with these metals.

Bullion

Buying platinum and/or palladium through bullion coins and/or bars is the safest way to do it, and it is appropriate if you are taking a long-term approach. Which you buy will be based on your budget and personal tastes. Obviously, platinum is much more expensive than palladium at this time so that will be a factor.

Fortunately, those who want platinum bullion do not have to buy a full ounce given the four-figure price tag. There are dealers that make platinum available in smaller sizes (such as a half-ounce and quarter-ounce). For more details on buying bullion, check out Chapter 10.

Futures

Platinum and palladium are widely traded in the futures market. You can speculate with futures contracts directly or (my preference) buy options on those futures contracts. The most suitable place to trade futures in the United States is the New York Mercantile Exchange (NYMEX). You can get full details about NYMEX and platinum/palladium futures at www.nymex.com. You find out more about futures in Chapter 14.

Platinum futures

A platinum futures contract is based on 50 troy ounces of platinum (trading symbol: PL). Therefore, at a per-ounce price of $1,270 times 50 ounces, that means a platinum futures contract would be valued at $63,500. A very hefty price indeed. When you trade futures contract (such as platinum), you don't have to have the full amount in your account; you can put a portion of the amount in your futures account (margin). How much? As an example, if you opened an account as of September 5, 200,7 and you wanted to trade one platinum futures contract, you'd have to put in a minimum of $2,700 (as a non-member customer).

With futures, you have the power (and potential problem) of leverage. If platinum goes up $50, then the futures contract's value would increase to $66,000 (50 ounces times the new price of $1,320 per ounce) and you could have a potential gain of $2,500. Since the margin in this case was $2,700, you would nearly double your money. But what if platinum goes down?

Keep in mind that if platinum goes down in price you may have to put more margin in your account. You would get the dreaded margin call from the broker to contribute either more money or marginable assets into the account. The volatility and risk make futures financially dangerous for those who are not experienced and with limited funds.

Palladium futures

A palladium futures contract is based on 100 troy ounces of palladium (trading symbol: PA). If palladium is at $350 per ounce, then the futures contract would be worth $35,000. This is certainly cheaper than the platinum futures contract but not exactly chicken feed. The same basic rules and risks would apply. Again, check out Chapter 14 regarding more details about the risks and opportunities in futures.

Both platinum and palladium futures (along with futures in general) are appropriate for investors and speculators who understand the risks and rewards of this fast-moving market. It is generally not appropriate for conservative or risk-adverse investors and it sure isn't a good place for your savings, your rent money, your nest egg, or what is in your kid's piggybank.

Final points

As a last reminder, options on these same futures contracts are a very active market and done at the same exchange (www.nymex.com). At this point, there are not "Mini" futures contracts on platinum or palladium but stay tuned. New derivatives are regularly issued and I am not surprised anymore. After all, if someone told me that after the brutal hurricane season of 2005 that we would have "hurricane futures," I would have called them crazy. But lo and behold we have hurricane futures. (Don't believe it? Check it out at the Chicago Mercantile Exchange's Web site www.cme.com!) What next? How about Election Season futures? Hmmmm . . . more wind.

Stock Investments

For investors, this is a more appropriate venue. There are mining stocks that are in the platinum group metals. Some examples of mining stocks are North American Palladium (symbol: PAL), Anglo American Platinum (symbol: AAUKD), and Impala Platinum (symbol: IMPUY). I list these only as examples so please don't take them as recommendations (confer with your stock advisor on whether they are appropriate for you).

It is important to point out that stocks, especially mining stocks, tend to be a "derivative" of the underlying asset. It is common to talk about an option being a derivative since an option "derives" its value from the underlying assets. The same point is generally true for futures contracts since futures (like options) are basically pieces of paper that have no intrinsic value; they are claims to things that do have value and certainly have value in the marketplace. Stocks can have the same quality. In addition, another trait that stocks have with futures and options is a certain measure of leverage, albeit to a different degree. Here's an example.

If I own platinum bullion and the price of platinum goes up 10%, then my bullion will go up a corresponding amount. But, all things being equal, a platinum mining stock will probably go up more than 10% (20% or 30% or more). Why? That type of mining stock (along with most mining stocks) probably has plenty of platinum on its property.

There is a very rough rule of thumb that says that a generic mining stock will have a 3-to-1 correlation in price movement compared to the underlying metal. In other words, if platinum goes up 10% then the stock of a strictly platinum-mining company would most likely go up three times as high (or more) because of the amount of platinum it has in the ground. It may have thousands or even millions of ounces but the important point is that a mining stock can be a leveraged play on that particular metal.

Keep in mind that few mining stocks are a "pure play" on a single metal. In most cases, the mining process can usually unearth many metals and minerals that can be clustered together in the same location. This is especially true with the platinum group metals. For more information on mining stocks, go to Chapter 12.

Mutual Funds and ETFs

There are very few mutual funds that exclusively get in investments (typically stocks) that mine the platinum group metals. The reason is that mining stocks in this singular area is a relatively small area of what is really a relatively small industry. As of the summer of 2007, the value of all the mining stocks in the U.S. stock market combined is still less than the market value of a single mega-cap stock (like Exxon Mobil or Microsoft).

Like base metals, they tend to be lumped in a mutual fund broadly referred to as "metals and mining," "basic materials," "raw materials," or even "natural resources." It is still worthwhile to do some research since it is not that difficult to find mutual funds with mining stocks. For more assistance on locating mutual funds with platinum group metals, check out the resources listed in Chapter 13 on mutual funds.

As of August 2007, there are no exchange-traded funds (ETFs) that are a pure play on platinum directly or the group in general. There has been ongoing chatter in the marketplace for a platinum ETF due to the metal's increasing price rise and popularity so it is possible that one could come out in the near future. Keep an eye on where ETFs typically emerge, the American Stock Exchange (www.amex.com).

Research Resources

To do your own research, here are the best resources and Web sites for the platinum group metals. Do some digging at the Web site since some of these sources have extensive information, data, and links on the mining industry in general but offer plenty of research on individual metals.

- Gold Fields Minerals Services (www.gfms.co.uk)
- Global Info Mine (www.infomine.com)
- Gold Sheet Links (www.goldsheetlinks.com)
- London Platinum and Palladium Market (www.lppm.co.uk)
- Metals Place (www.metalsplace.com; a news site for metals)
- Platinum Metals Review (www.platinummetalsreview.com)
- Platinum Today (www.platinum.matthey.com)
- United States Geological Survey (www.minerals.usgs.gov)

Chapter 8

Uranium

I'll try to keep the comments about "glowing returns" and a "radio-active future" to a minimum. But really, I think that uranium deserves consideration for any portfolio because it will be (and already is) a major player in the energy sector.

Maybe the first question you will ask is, "Uranium? Does it really belong in a book on precious metals?" The answer is a resounding "Why not?" It is indeed metallic, it does need to be mined, and it is indeed very precious. In the big picture, uranium is (and will be) playing an important role in the world as we enter the "twilight" period in our historic usage of oil (more about oil later in this chapter). Why address oil in a chapter on uranium? Uranium's future is tied to the prospects of oil in a big way.

Uranium is the primary element needed in nuclear power. Because there are different grades of uranium, you will hear a reference to "U3O8" yellowcake uranium. This reference is actually to the element symbol for uranium oxide concentrate. This is the primary grade used by industry and by all means don't confuse "yellowcake" uranium with any other cake.

In this chapter, I give you all the info you need to make smart decisions regarding investments in uranium so you can have your (yellow)cake and eat it, too.

Controversial Past, Bright Future

If any metal qualifies as going from an old beast to a new beauty, then uranium fits the bill. It was vilified stuff in the anti-nuclear protest days of the 1970s and 1980s. Yet in recent years environmental activists embraced nuclear power. Talk about a turnaround! Why this change of heart? It happened due to two major developments:

- **Nuclear power is safer than it used to be.** New technology coupled with better, safer, and more efficient ways to manage nuclear power and nuclear waste byproducts has made it a friendlier energy source.

- **The global warming controversy.** Public debate stirred up concerns over "greenhouse gases" and emissions from oil and coal based energy usage. Although there is considerable debate on the topic, that hasn't stopped political, business, and social movement towards energy usage that didn't impact the earth's climate. Nuclear power fits the bill nicely since nuclear power plants have no carbon emissions.

The uranium market

In terms of usage, uranium does have a "Dr. Jekyll and Mr. Hyde" quality about it. Uranium has both peaceful uses (energy and industrial applications) and maybe-not-so-peaceful uses (military applications).

Uranium's primary use in industry is as the fuel in nuclear power reactors to generate electricity. A secondary use is in the manufacturing of radio isotopes in medical applications. Uranium helps to provide about 16% of the world's electricity. As of the summer of 2007, there are 443 active commercial nuclear power plants operating in 30 countries across the globe providing 370 gigawatts of electricity. Anywhere from 50 to 200 new nuclear power plants are in some stage of planning worldwide during the next 10 to 15 years.

Demand and usage from individual countries are very diverse. France doesn't have much in the way of uranium but it imports much since 70% if its energy needs is provided by nuclear power. Australia, on the other hand, doesn't use very much uranium so it is a major exporter of uranium. China has a voracious need for energy for its growing economy so it's planning to significantly expand its nuclear power (adding 30 to 40 power plants during the next five to ten years). How about us?

As of 2006, the United States derives about 19 percent of its total electricity from 66 nuclear power plants. There hasn't been a new nuclear power built since the late 1970s although that may change due to the new realities for our economy, such as peak oil.

Uranium supply

Part of the bullish picture for uranium is the limited supply of it. However, the supply is more limited than you think.

Uranium is found in areas that are often geologically fragile. This means that problems may occur when companies attempt to extract it. Although problems with mines are present in all types of mining, uranium mining is especially delicate. Two recent major mining problems illustrate this.

In September 2006, flooding occurred at the Cigar Lake mine due to geological instability. The mine is owned and managed by the large mining company Cameco (listed in the New York Stock Exchange with the symbol CCJ). This was a very significant event for the uranium industry since the flood shut down production of a mine that geologists report has the world's largest deposit of undeveloped commercial-grade uranium. This mine was scheduled to reach production level by 2008 but now it may take much longer. Although the company is progressing with efforts to remedy the project, the earliest that production can begin is 2010 or 2011. Since Cigar Lake had estimated reserves of over 226 million pounds of U_3O_8 grade uranium, this is a major setback for available, worldwide supply.

If that wasn't bad enough, then there is another problem in Australia; the Ranger mine incident. Tropical Cyclone George flooded the Ranger mine, which was already a producing property. The drop in production was estimated at about 4 million pounds. It is likely that other problems with uranium mines may surface. All this adds up to a big negative on the supply side of the uranium market. How about demand?

Uranium demand

Because of the world's voracious need for energy, this bodes well for uranium. China, Russia, India, Japan, South Korea, and even the United States have major plans to build more nuclear power plants. Just China and Russia will add 80 to 100 nuclear power plants by the end of the next decade. Until uranium production catches up, this burgeoning demand will be met by a combination of current production and what is available in stock piles. The stock piles include uranium being recovered and re-used from sources such as missile warheads formerly used by the old Soviet Union.

When you combine limited or shrinking supply with strong and growing demand, you have an investment winner. Uranium radiates long-term success!

Uranium's market performance

The price of uranium experienced a strong bull market in the late 1970s before disenchantment with it sent the price plummeting in the early 1980s. Uranium (U3O8 grade) witnessed its spot price rise from under $7 a pound in 1972 to just over $40 by 1979. Keep in mind that the demand for uranium was not just because of the energy situation; it was also due to the military buildup of nuclear forces by the United States and the old U.S.S.R.

By 1980, uranium started on a long, steady decline before it hit rock bottom in 2000 (back below $7 a pound). This bear market lasted two decades until world demand caught up and surpassed world supply. So as with most commodities in this decade, it turned around in 2000 and started an impressive and uninterrupted climb to $138 a pound by the spring of 2007. During the summer of 2007 it experienced a sizable correction and retreated to the $90 level by early September 2007 (see Figure 8-1).

Figure 8-1:
Uranium
chart
1972–2007.

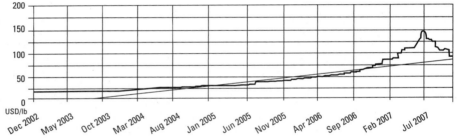

Some of the major reasons for uranium's price rise during 2000-2007 are the reduction in above-ground supplies available, the increased costs of oil, and the outlook for greater demand. The correction in the summer of 2007 occurred as the market perceived reduced demand for energy due to a slowing U.S. economy and seasonal weakness. (Summer time is usually a slow or weak time for commodities in general.) Even the hottest market can have pull-backs in the price which is actually normal and necessary for a long-term bull market.

A major factor going for uranium is the imbalance in supply and demand. Much of the demand is due to the rising price of oil. However, demand can grow faster than supply because it takes a long time to find and mine new sources of uranium. A uranium company has to jump through many hoops (permits, regulations, political barriers, and so on.) over a period of years before it can stick the first shovel in the ground. This is an important fact for nimble investors and speculators.

That '70s metal

With the looming energy crisis of the early 1970s upon us, America sought alternatives. Because the United States hit "peak oil" circa 1971, the need for alternative energy became apparent. More about peak oil in the next section. Anyway, the energy crisis of the 1970s was a primary catalyst for oil-consuming countries such as the United States to diversify into other energy sources. This in turn helped to create a boom for nuclear power during the late 1970s. However, it was not without controversy. The potential hazards of nuclear power, both real and perceived, dogged it into a national controversy.

Moves like "China Syndrome" and incidents such as the 1979 accident at the Three Mile Island nuclear power plant near Harrisburg, Pennsylvania, turned public (and hence political) opinion against nuclear power.

Peak oil and uranium

I use an example in the gold chapter about umbrellas and rain. I compare gold to an umbrella and monetary inflation to rain. When you see the problem of rain (monetary inflation), you can see the solution of an umbrella (gold). The same umbrella-and-rain example fits with uranium and . . . peak oil. I mention it earlier, but it is something that you should be aware of.

Peak oil is a condition first covered in landmark research done by the geologist Marion King Hubbard in the early 1960s. Peak oil refers to the lifespan of a pool of oil found in the ground. The first 50% of its life, the oil comes gushing out easily. The second 50% is more difficult and expensive to extract. Peak oil doesn't necessarily mean we are running out of oil; it means that we are running out of cheap oil. This condition will have a profound effect on our economy and lifestyle. In 2007, the 20 largest oil fields in the world are declining in production and in the coming years we will need to find new sources of energy. (Got uranium?)

To find out more about peak oil, you can go to the Association for the Study of Peak Oil and gas (www.aspo.net) and there are more links to resources on peak oil through the Web sites below. An excellent book on the topic is *Twilight in the Desert* by Matthew Simmons (published by Wiley). Consider peak oil as a "must" topic in your research.

Uranium Investing Vehicles

First let me say right up front that you shouldn't have uranium in any physical form (not suitable for kids, seniors, or as gift items)! Did I really need to say that? Considering how hare-brained things are these days, why not? There are bottles of sleeping pills that have a notice on them that says "Warning! The contents may cause drowsiness." Huh? Great . . . just as I was about to operate heavy machinery! Anyway, I can look like a consumer hero and tell you to stay away from uranium in all its physical forms. But embrace it as an investing opportunity. The following sections give you some of the ways that you can power up your portfolio returns with uranium.

Futures

Uranium in the form of a futures contract is a very new animal. Uranium started trading in the futures market on the New York Mercantile Exchange (NYMEX) in May 2007 with the symbol "UX." The futures contract is attached to 250 lbs. of U3O8 grade uranium and it is traded "cash only." In other words, all trades are settled in cash; there is no delivery of actual uranium allowed (whew!). Actually, if anyone was looking for physical delivery we can only hope that the exchange lets those national security folks know about it.

Because uranium futures are relatively new, the market does not (yet) have options on futures but there's no reason why that can't change in the near future. As the market heats up (pardon the pun), more activity and excitement will result in more ways to play uranium.

Mining stocks

Generally the same issues with mining stocks apply to uranium stocks so Chapter 12 is important to read. Uranium stocks have the same issues of volatility as the regular metals mining stocks. However, there are different factors affecting uranium stocks.

If you recall from prior chapters, inflation is a major driving force for precious metals such as gold and silver, and therefore it has a major influence on gold and silver mining stocks. For uranium stocks, the more direct influence is, in a word, "oil." In other words, the world's consumption and demand for energy (oil and otherwise) is the primary driver for uranium stocks. As the price of oil in this decade soared from under $10 a barrel to well over $70, uranium also went higher and more steeply; it went from under $7 a pound to over $90 in the same period. How did uranium stocks perform?

As you can guess, uranium stocks went up dramatically and beat the pants off the general stock market. Both classes of uranium stocks, the major producers and the junior explorers, did exceptionally well. Take Cameco, for example. It went from under $2 per share in 2000 to over $40 in 2007. (That does not include two stock splits during that time, which means you did even better.) Some of the junior uranium stocks did even better.

We can see the rewards with uranium stocks but the risks do come along as well. A well-selected uranium stock has powerful profit potential in this environment but keep in mind the potential problems. In addition to having the usual challenges by mining firms (regulations, political risks, and so on) there are the geological risks as well. Remember Cameco's Cigar Lake problem. When this incident hit the news, Cameco's stock fell from about $39 to $31 in about four weeks. Fortunately, it is a well-financed company that was able to get through this setback, but what if it were a junior exploration firm? It would have gone into bankruptcy and the share price would have been vaporized.

The bottom line is that the larger uranium stocks are suitable for long-term, growth-oriented investors. The junior mining stocks are more speculative so they are more suitable for aggressive investors. For more research into uranium stocks, see the resources at the end of this chapter.

Uranium exchange-traded funds

For investors, investing in uranium through ETFs is the safest way to go. Although there are currently only two choices, these are in fact great choices. Uranium ETFs remove the risk of holding only one company. There is risk with uranium ETFs but the risk is related to the overall industry. If the industry goes down (meaning market value), then the ETF will go down; if the industry prospers and prices go up, the ETF will rise in value as well. Since uranium is a worldwide market and its prospects look good for the next few years, uranium ETFs are a solid way to play this market.

Uranium Participation Corporation

The Uranium Participation Corporation is listed on the Toronto Stock Exchange (www.tsx.com) with the symbol "U" and it's been around since 2005. Fortunately for U.S. investors, the security is also listed on Nasdaq's Over-the-Counter Bulletin Board (www.otcbb.com) with the symbol URPTF.

URPTF is managed by Denison Mines and it is a way for investors to easily participate in uranium without having to buy stock in a particular company. The security mirrors the market price of uranium directly. Technically, it is not an ETF but it has enough of the characteristics to effectively be one.

As an over-the-counter security, there are no options on it and you can not buy it on margin in the U.S. Consider it an aggressive choice for your portfolio.

The Market Vectors Nuclear ETF

The Market Vectors Nuclear Energy ETF (symbol: NLR) is an ETF issued by Van Eck Global. It mirrors a nuclear energy index called the DAX Global Nuclear Energy Index (DXNE). The DXNE is an index that tracks large uranium mining companies (and large firms related to uranium such as processing, services, and so on). Currently, they track 38 companies engaged in the nuclear energy industry and that are traded on leading global stock exchanges.

For long-term investors, this is a solid way to invest in the uranium market. You get the safety of diversification (as if it were a mutual fund) and the convenience of an ETF. More details on ETFs are in Chapter 13.

Other choices

As I write this book, more choices are jumping on the bandwagon. Premiering on the Toronto Stock Exchange is the Uranium Focused Energy Fund, which invests in securities of issuers that operate in or have exposure to the uranium sector (symbol: UF.UN on the Toronto Stock Exchange). As the uranium market heats up (sorry about that), you can expect more choices for investors and speculators.

Options

Of course, those who want to get aggressive and "push the envelope" for more speculative gains, options are the way to go. Right now, the most active options areas for uranium are options on the larger uranium companies. As of the date of this book, there are no options on uranium futures or on uranium ETFs but I expect that to change. More "options on options" are probably on the way.

Fortunately, there are many choices on options with the major uranium companies and as of September 2007, there are options with expiration dates going out to 2010. Another point to keep in mind is that many smaller uranium stocks do have warrants which are very similar to options and also have expiration dates that go out two years or longer.

For options, do your research by getting the stock symbols for the major firms and find out if there are options available at the Chicago Board Options Exchange (www.cboe.com). For warrant information, check with the individual companies. Use the resources in the next section to find companies and more information. Happy digging!

Resources

For information on investing in uranium, try the following:

- Trade Tech Uranium Information Web site (www.Uranium.info)
- Uranium Seek (www.uraniumseek.com)
- UXC Consulting (www.uxc.com)
- Uranium Information Center (www.uic.com.au)
- World Information Service on Energy (www.wise-uranium.org)
- Energy Information Administration (Official Energy Statistics of the U.S. Government; www.eia.doe.gov/fuelnuclear.html)
- Global Info Mine (www.infomine.com)
- World Nuclear Association (www.world-nuclear.org)

Chapter 9

Base Metals

O kay. Admit it. As soon as you got this book you immediately flipped to this chapter. Didn't you? After all, who wouldn't want to read about copper and nickel? Well, the chapter is a reality and now you don't have to wait for the movie! But just in case you count base metals instead of sheep at night to get to sleep, I'll keep the technical stuff to a minimum. You can thank me later.

Anyway, base metals can be exciting as a speculative and investment vehicle because they're indeed the building blocks of civilization. Putting your money into base metals investments is a bet on society at large and its continual advancement. Whatever of substance is to be created — cars, computers, buildings, highways, and so forth — needs base metals.

Now, to add some interest in this chapter (as if I needed to with this scintillant topic), I will quiz you with the following question: What base metal was used the most by the Swiss navy? The answer appears somewhere later in this chapter. That aside, use this chapter to check out what profitable opportunities await you (and why) with base metals.

Understanding How Base Metals Fit into Your Portfolio

You've probably never been more interested in base metals than you are now, or perhaps you really aren't interested at all and you just flipped to this chapter by accident. In either case, you need to check out the following sections so you can fully understand why base metals should play a role in your portfolio.

The building blocks of society

The world's population keeps growing. In recent years it has surpassed 6 billion and in years to come there will be more and more people. Now whether or not you decide to become a hermit, you can at least benefit from this relentless growth in "people" production. After all, more people means that more things have to be created. More people mean more cars, more houses, more hospitals and schools, and oodles of other stuff that need raw materials.

A huge part of the world's need for more base metals is the economic growth of huge population centers such as China, India, and lesser (but still robust) countries such as Brazil. In recent years, China and India transformed themselves into economic giants as they became more free-market in their orientation. This culminated in voracious demand for consumer and industrial goods. So, current and future demand for base metals becomes assured.

The need for base metals as the building blocks of society isn't just limited to the relatively new economies burgeoning across the globe; it is also an issue in the United States and Europe. Our relatively older societies have massive infrastructures that need replacement or refurbishing. The old building blocks need to be replaced with new building blocks. In other words, base metals will be in demand for the foreseeable future.

As populations and economies keep growing, base metals such as copper, aluminum, zinc, nickel, and lesser-known metals will need to be provided. Well-run companies that find these metals and extract them from the earth will continue to be profitable (for stock investors as well). Trading and speculating in base metals will continue to provide opportunities for us in futures (and options on futures).

The up and down for base metals

The first thing that you would probably extract from the above section is that supply and demand for base metals are definitely positive. Economic growth is a huge plus for base metals and investments tied to them. Long-term investors and speculators will ultimately see growth in their base metals–related investments. But base metals are a double-edged sword (they are probably in the sword, too). Because the fortunes of base metals are tied to the economy, they will therefore be very sensitive to economic shifts.

Base metals and inflation

Besides supply and demand, another factor in the ongoing picture for base metals is inflation. Why is that a factor? Review the following two points and put your economic thinking hat on. *Inflation* is the excessive creation of a country's currency. As in the case of the U.S., the more dollars the government creates, the more dollars there are out there chasing a limited supply of goods and services. Inflating the currency results in price inflation. How does this impact base metals?

Base metals are a finite commodity. In other words, you can't make any more of them. Whatever base metals you find (both above and below ground) . . . that's it! As a strictly limited resource, you can't simply grow more (as we can with grains or other food stuffs), and you can't create more (like those pesky widgets we keep reading about). You get the picture. So that's the first point: whatever you have . . . that's it.

The supply of humans and fiat currencies (such as dollars, yen, and so on) keep expanding. The bottom line is that more humans with more money means more demand, which is inflationary and results in higher prices for finite supplies of stuff, such as base metals (and other metals mentioned throughout the book). Hence, the price of base metals–related investments has a bright, long-term picture. Kapeesh?

Base metals versus precious metals

Some investors may have the attitude that "metals are metals" and that they have the same economic and financial qualities, but that is not so. Base metals and precious metals, despite the "metallic similarity," are quite different animals. It is like talking about the jaguar in the Amazon jungle and the Jaguar in the dealer showroom (actually the one in the showroom has both base and precious metals but that's a different story altogether).

The performance of base metals as an investment category is closely tied to the performance of the general economy, especially in cyclical industries such as housing, construction, technology, and automotive. In addition, base metals are used in industries tied to government spending that may or may not correlate with economic growth. Good examples of this are national defense and municipal infrastructure expenditures. Tanks, bullets, and bridges use lots of base metals.

Precious metals, on the other hand, have different dynamics. Their historical performance bears this out. When inflation and economic stagnation (*stagflation*) were raging in the 1970s, precious metals provided superlative gains for investors that far outpaced base metals and for obvious reasons.

Let me just give you a down-'n'-dirty, nitty-gritty comparison: When the economy is wonderful and growing, consider base metals. When the economy is struggling and uncertain, consider precious metals. When energy prices are soaring, precious metals (gold, silver, and . . . yes . . . uranium) are a good consideration. In real time (as I am writing this book), I think that a prudent course is to have "all the above." In other words, diversify.

Past performance

Imagine a greater percentage that some of the sexy, big-name tech stocks couldn't outdo. Yes, if base metals were in Hollywood they would be like Phyllis Diller, but the returns on base metals would be like Sophia Loren!

During 2000–2007, most of the attention was on the general stock market, real estate, and energy. Few noticed precious metals and even fewer noticed base metals. Yet, base metals as an investment category performed very well. GFMS Metals Consulting (www.gfms-metalsconsulting.com) created an index to track the general price performance for the leading base metals (on the official LME cash settlement price for primary aluminum, copper, lead, nickel, tin, and zinc) based on the official London Mercantile Exchange (LME) cash settlement prices extracted from daily trading activity on the LME. The GFMS index is an average of the six base metals prices with equal weight given to each of them.

Covering All the Bases

A base metal is a common and generally inexpensive metal that can oxidize or corrode relatively easily (not that that is a selling point, mind you). Base metals are also referred to as nonferrous metals, which means that they are metals or alloys that are entirely (or mostly) free of iron in their physical makeup. These qualities distinguish themselves from precious metals that generally do not corrode (such as gold or platinum, which are referred to as *noble metals*).

Centuries ago, base metals were used in alchemy to try to convert them to precious metals (but . . . surprisingly . . . to no avail). Today we don't try to convert base metals to precious metals since base metals had great use in more down-to-earth pursuits like building houses and such. Besides, if you want precious metals that badly then just go and buy them instead. In the following sections, I give you the stars of this show — the most common base metals that offer investment opportunities.

Copper

Copper has been used by civilization for centuries. Few commodities are as "in lock step" with the world economy as copper. After iron and aluminum, it is third on the hit parade of the world's most widely used metals. It is used extensively in industries such as construction, manufacturing, and machinery parts. As the rest of the world modernizes and keeps building infrastructure, copper will continue to have a bright future. How has it done so far? Just check out Figure 9-1.

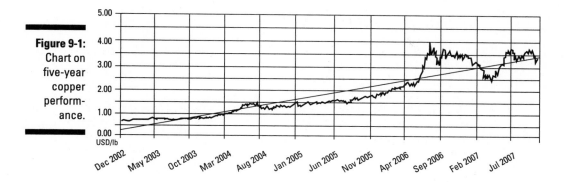

Figure 9-1:
Chart on
five-year
copper
perform-
ance.

In September 2002, you could have bought copper for about 60¢ a pound. Five years later, a pound of copper is priced at $3.25 for a total percentage gain of over 441%. Sweet!

Keep in mind that the above chart (and the charts that follow) shouldn't just show you how well that particular base metal did over the long term (such as five years) but also to show how the short term (such as a three-month period) can be so volatile. Look at the price of copper during 2006. In the first six months it shot upward like a rocket ship, doubling from $2 in January 2007 to $4 by May 2006. It then experienced a mini-crash as it plunged to under $2.50 by February 2007 (ouch!).

There are lessons here for investors, traders, and speculators of all stripes. The long term is easier to manage than the short term since the overall trend will benefit you in spite of the zigs and zags. Short-term investors can get burned since corrections can come fast and furiously. Traders, on the other hand, look for this volatility so that they can put on a variety of positions that could benefit from the downs as well as the ups.

Aluminum

Aluminum is a light, corrosion-resistant metal widely used in our economy. And I don't just mean the aluminum foil you keep in your kitchen drawer! You will also see it used in aerospace, housing, automotive, and technological applications.

Believe it or not, the largest single area that consumes and uses aluminum is the transportation sector. It absorbs approximately 30% of U.S. aluminum production. The uses for it keep growing and so it is a good investment bet as the world economy keeps growing as indicated in Figure 9-2.

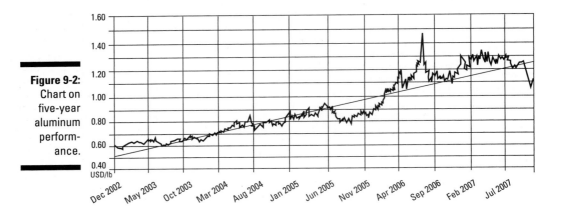

Figure 9-2:
Chart on
five-year
aluminum
perform-
ance.

Aluminum was 60¢ a pound in September 2002 and after briefly spiking to nearly $1.50 in May 2006 still settled at $1.10 per pound by September 2007 for a total percentage gain of 83%. Considering how some of the other base metals did over the same time frame (the other primary base metals went up from 288% to 650%), that 83% seems like a real laggard! Then again, when you compare it to standard barometers of investment performance (such as the S&P 500 index), it's a solid gain. Incidentally, the S&P 500 went up 65%.

Nickel

Nickel is used in many applications as an alloying metal. An alloy is a material composed of two or more chemical elements of which at least one is a metal. Stainless steel, for example, is an alloy. It is composed of carbon, chromium, iron, and . . . you guessed it . . . nickel.

Nickel has a high melting point and it is resistant to corrosion. Early in the 20th century, it was discovered that by combining nickel with steel, even in small quantities, the durability of the steel increased significantly with regards to corrosion resistance and strength. The steel industry is now the largest consumer of nickel. Figure 9-3 shows you nickel's performance.

In the five-year span that ended September 2007, nickel's price went from under $3 a pound to $12.50 for a total percentage gain of 317%. Pretty good but that's not the whole story. In 2006, nickel was the single highest performing asset. In fact, during the 12-month period ending in June 2007, nickel went from $5 to $25 in one year . . . a blistering 400% gain that would make any trader salivate. Of course, in the ebb and flow of the market, nickel took a breather and retreated to $13 by August 2007. What a ride!

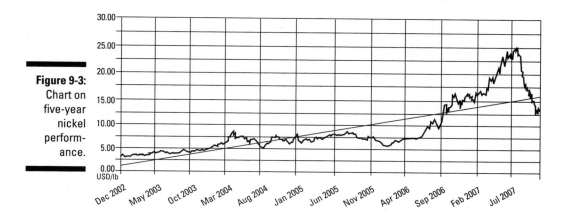

Figure 9-3:
Chart on
five-year
nickel
perform-
ance.

Zinc

Zinc is usually found as a co-product with other base metals such as lead. It is mainly used for industrial purposes such as galvanizing which gives extra protection against corrosion in a wide array of manufactured goods such as building construction, automobiles, machinery and other hard goods. As the world economy continues to grow, the price of zinc will certainly grow along with it (see Figure 9-4). Keep this in mind next time you have a zinc lozenge.

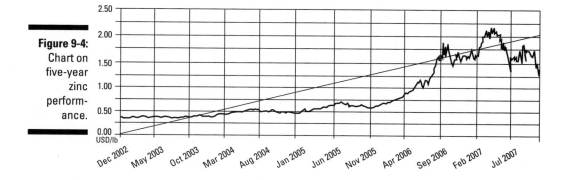

Figure 9-4:
Chart on
five-year
zinc
perform-
ance.

Lead

The uses of lead can be traced back to the earliest days of history. The Romans described it as the basest of base metals due to the ease with which it could be beaten or melted. In recent time, lead has come into its own with the evolution of applications such as petrol additives, pigments, chemicals, crystal glass, and batteries. The largest market for lead is in battery production, consuming approximately two thirds of the lead produced in the western world. Lead's relationship with the LME began before the turn of the century, although dealings were unofficial. By 1903, it was being traded in a small secondary ring, but still without an official price setting. Lead was first officially traded in 1920 and the current standard lead contract, introduced in 1953, has become the mainstay for international free trade in this metal. You can check out lead's performance in Figure 9-5.

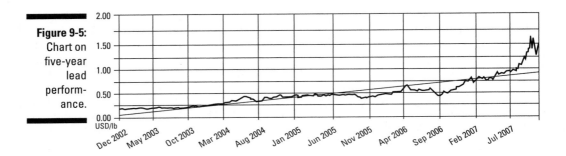

Figure 9-5: Chart on five-year lead performance.

Whew! What a performance. When they say "get the lead out" they're not kidding. During the five-year time frame, lead went from 20¢ a pound to $1.50 for a total percentage gain of 650%. It just makes the tears well up in my eyes!

Tin

Tin is a base metal that has been fused to other metals such as lead, zinc, and steel to prevent corrosion. Tin-plated steel containers are widely used for food preservation, and this forms a large part of the demand for tin. Tin's performance is illustrated in Figure 9-6.

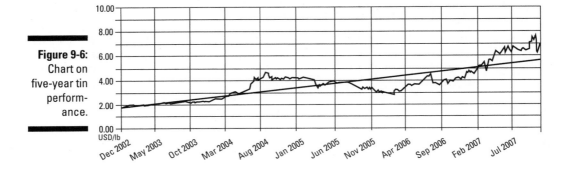

Figure 9-6: Chart on five-year tin performance.

As you can see, tin has performed very well in recent years. The price of tin was about $1.80 per pound in September 2002 and it hit $7 by September 2007. That makes for a total percentage gain of nearly 289%. If the Tin Man in the Wizard of Oz only knew! His tin futures could have bought him a heart and a nice duplex in Kansas.

Other metals

How about some other metals that few are talking about? As if there is a cocktail party where some fellow with an ascot and brandy is saying, "Hmmm. You mean there is more to life than merely lead and tin?!" Get a life. This stuff doesn't have to be interesting — it just has to be profitable — which leads us to other metals that may be obscure but still offer profitable opportunities.

Using the resources at the end of the chapter, take a look at metals that have stirred some interest in the marketplace. You just might find the next metal to score a triple-digit percentage gain in a short period of time. Here are some potential winners:

- ✔ Molybdenum
- ✔ Cobalt
- ✔ Titanium
- ✔ Manganese
- ✔ Chromium
- ✔ Iron ore
- ✔ Tungsten
- ✔ Kryptonite

Base Metal Investing Vehicles

The ways to invest, trade, or speculate in base metals are really quite varied. Whether you are aggressive or conservative in your approach, whether you are long-term or short-term in your time horizon, there are a variety of ways to include base metals in your portfolio.

Futures

Base metals in the form of futures contracts are the most speculative vehicles. Futures, in general, tend to be risky and volatile and metals-related futures (either precious or base) tend to have even greater risk and volatility.

Experienced traders and other short-term speculators embrace this roller-coaster aspect of the futures market. For anyone else, either have extra antacid and aspirin handy or just plain avoid this area.

For those who are aware of the risk and still want to proceed with a portion of their risk capital (as opposed to rent money or your retirement nest egg), then base metals futures offer lots of choices and opportunities.

To investigate what base metals futures contracts are available, the best places to start your research are at the New York Mercantile Exchange (www.nymex.com) and the London Mercantile Exchange (www.lme.com). The futures market is covered in Chapter 14.

For those who want to try risking some money on futures, it is probably more advisable to look at options as a way to limit risk and still have opportunities for speculative gains.

Options

In recent years, the variety and availability of options has been explosive. Options (such as calls and puts) are a huge and popular segment of the world of derivatives. I couldn't do justice to this topic in this segment so please read Chapter 15 for a better treatment of this phenomenal topic.

With options, I can choose strategies that can be ultra-aggressive or very conservative. You can use options to make a bet on the underlying investments to go up . . . or down. You can use options to make potentially explosive gains or steady income. Some options strategies can be risky and volatile; other strategies can be used to increase safety and peace of mind by helping to limit risk. The options strategies available for base metals (and their related investments) are just as varied as in any other assets.

There are options on

- Futures (all the primary base metals mentioned in this chapter have options available on them)
- Stocks of base metals–mining and –processing companies
- Base metals–related ETFs

Options are also good vehicles that can work in tandem with your investments. If, for example, you have shares of stock in a major base metals–mining company, check to see if it publicly traded options on the stock.

You can find out easily by going to the Chicago Board Options Exchange (www.cboe.com) and get a detailed listing of options available by going to their Web site and entering your stock's symbol. Then you employ strategies that could generate extra income from your stock holdings (such as through covered call-writing). Or you can do a protective put to limit your stock's downside risk. These strategies are further detailed in Chapter 15.

Mining stocks

For most folks, especially those who have a long-term investing approach (again, as opposed to short-term trading and/or speculating), base metals–mining stocks would make sense. These can be easily broken down between small-cap and large-cap stocks. The *cap* is a reference to market capitalization (not headgear). *Market capitalization* is just a fancy phrase for the company's market value. The market value is simple math: the number of shares outstanding times the current market price per share. A company that has 1 million shares outstanding and is $12 per share would then have a market cap of $12 million. Easy! Market caps of under $1 billion are generally considered small cap. One to ten billion is generally the range for mid-cap, and above $10 billion is large-cap. If they are, say, over $50 billion then they are referred to as mega-cap. Just to jump to the other end, firms that are under $100 million tend to be called "micro caps." Don't be too hung up on the exact numbers; I think you get the picture.

Base metals miners tend to be broken up into firms that are explorers and the majors. The explorers are typically small caps and are also called junior mining stocks. The majors are established firms that are producing and selling what metals and minerals that they are extracting from the ground.

Fundamental analysis

Because I think that the stocks should be considered long-term, fundamental analysis is the preferred approach (versus technical analysis, which is covered in Chapter 18). In fundamental analysis, you look at the company's financial strengths, profitability, and its standing in the marketplace.

Investing in stocks and other investment vehicles that cater to human needs is a long-term winning formula. Since base metals are so necessary for a modern society it stands to reason that if you have a company that consistently provides it at a profit and that manages its financial affairs accordingly, you have a solid stock pick. Fundamental analysis brings it together: A profitable company that provides something that the market wants. But let's dig a little deeper.

There are very few companies in the metals industry that only do one metal. Many metals are found in the same geologic locations. Silver, for example, is usually a byproduct that accompanies other metals such as gold or copper. Therefore, it is common that mining companies have more than one base metal (and/or precious metal) that they mine and resell. This is a net positive since companies with diversified mining assets are a safer bet than a company that is banking on a lone metal.

Keep in mind that the resources at the end of this chapter are also great resources to find and analyze mining companies. Lastly, the mining company's leading asset is, of course, its proven reserves of metals. In that case, a mining firm that specializes in base metals has similar attributes to precious metals mining firms. Get some important points and resources about choosing mining stocks in Chapter 12.

Looking at the numbers

I won't make you go nuts by looking at and analyzing all the numbers, but you should ask some crucial questions about the financials of the company after you get past the point of knowing that the company offers something that the market wants and needs:

- **Is the company profitable?** It should be consistently making a profit in each of the past three years and each successive year's profit should be greater than the year before. This data is in the company's income statement that is made publicly available. If you are looking at junior mining stocks which may not be generating a profit, then you should look into the company's cash flow and cash on-hand to keep it afloat until it scores with metals discoveries.

- **Does the company have enough capital on hand to finance growth?** You can look at the balance sheet and see that it has cash on hand, accounts receivables (money coming in from customers), and lines of credit. Mining is an expensive pursuit so adequate capital is necessary as it takes time to extract saleable metals from the ground.

- **Does the company have low or manageable levels of debt?** Its assets should be double the size of its liabilities (or better). Having too much debt will be a problem, especially if there is a slowdown in the economy which affects the company's cash flow (and therefore its ability to pay its debts).

There's more to choosing a company than that but those are among the top considerations. Get more details on the financials of the company with some guidance and resources in Chapter 12.

Mutual funds

There are very few ways to invest in base metals through mutual funds and exchange-traded funds (ETFs; see next section) that are "pure plays" (being 100% in base metals or close enough) in base metals. Typically, mutual funds that have base metals-related investments (such as mining stocks) would be in a more broadly based sector called basic materials, raw materials, or natural resources.

Additionally, there are mutual funds that cover metals in general. Some tend to lump in base metals with precious metals to have an "all-purpose metals sector fund." Although this sounds like it makes sense it is really an optimal strategy since base metals work a little differently compared to precious metals. To play base metals based on expectations of economic growth, they are best paired with natural resources with similar supply and demand dynamics (such as lumber and cement). Precious metals tend to play off of different dynamics such as inflation and economic uncertainty.

When you seek out mutual funds that generally cover the "natural resources" (or commodities) industry you will see plenty of choices since many mutual fund firms have added mutual funds that emphasize this industry since the start of the natural resources bull market in 2000.

 Because mutual funds that concentrate on natural resources can have a wide variety of choices (both stocks of companies that do base metals and many that don't) it is advisable to check the most recent "portfolio snapshot" provided by the mutual fund manager. It usually accompanies the mutual fund prospectus and it is commonly available at the firm's Web site. In addition, many major financial Web sites (such as www.marketwatch.com and www.finance.yahoo.com) will usually list the top holdings of that particular fund.

Exchange-traded funds (ETFs)

Alas, although the ETF choices are limited for base metals enthusiasts, the growing interest in this necessary industry will no doubt offer more alternatives in the coming months and years. After all, there were no base metals-related ETFs at all in early 2006. But by January 2007, there were two with substantial base metals exposure (see below). To see what else may become available, check in every now and then with the exchange that has been leading the pack with ETF listings, the American Stock Exchange (www.amex.com).

SPDR Metals and Mining ETF (XME)

XME is an ETF issued by Standard and seeks investment results that, before expenses, generally correspond to the performance of the SandP Metals and Mining Select Industry Index.

How has it fared since it started trading (June 22, 2006)? That day, XME closed at the price of $44.85 per share. At the end of trading on August 31, 2007, it closed at $57.92. For the time frame (a hair over fourteen months), it did well with a total percentage gain of 29% (not bad!). During the same time frame, the S&P 500 index scored a total percentage gain of 18.3%, which is pretty good too but again, base metals displayed their "precious mettle."

The PowerShares DB Base Metals Fund (DBB)

DBB is an ETF issued by DB Commodity Services. They maintain a series of ETFs called "PowerShares TM." DBB premiered in January 2007 and it tracks the Deutsche Bank Liquid Commodity Index–Optimum Yield Base Metals Excess Return (whew . . . that's a mouthful). It is an index ETF composed of futures contracts on three of the most liquid and widely used base metals – aluminum, copper, and zinc. DBB is intended to mirror the performance of the base metals (also called "industrial metals") industry. How has it fared?

On its first day of trading (January 5, 2007), it closed that day at $23.71 per share. On the last trading day of August (barely eight months later) it ended the day at $25.93. A not-too-shabby percentage gain of 9.4% (for a measly eight months, that's more-than-respectable). During the same time frame, the S&P 500 (a broad stock market index) had a percentage gain of about 4.6%. In other words, DBB did twice as well as the S&P 500 index. Keep in mind that the time frame is too short to make a valid observation. However, it does give a small example of the power of diversifying your portfolio with some of the building blocks of society.

For now, there are some great ETFs that offer a conservative way to add base metals exposure to any portfolio. More information on mutual funds and ETFs in Chapter 13.

By the way, about that question from the beginning of the chapter, you know, the one about what base metal was used most by the Swiss navy. The answer is . . . fooled you! There is no Swiss navy (it is a land-locked nation). However, you do earn extra points if you said that base metals are used in their watches and army knives.

Base Metal Resources

Seriously, as the world uses up more and more base metals, far-sighted and informed investors will gain. Hopefully, this chapter gives you a fair glimpse into the opportunities and pitfalls of base metals. Continue your research, starting at the following great places:

- United States Geological Survey (www.usgs.gov)
- Base Metals (www.basemetals.com)
- GFMS Consulting (www.gfms-metalsconsulting.com)
- Metal Prices (www.metalprices.com)
- London Mercantile Exchange (www.lme.com)
- About.com's Metals section (www.metals.about.com)
- Kitco Base Metals Data (www.kitcometals.com)
- Global Info Mine (www.infomine.com)

Part III:
Investing Vehicles

The 5th Wave   By Rich Tennant

"Precious metals? ETFs? Mutual funds? I say we stick the money in the ground like always, and then feed this guy to the sharks."

In this part . . .

*W*hile Part II gives you the "what," this part gives you
the "how." How can you proceed as an investor,
speculator, or trader in the world of precious metals? The
vehicles range from physical bullion and collectibles to
stocks, futures, and even mutual funds. Discover more
about the cornucopia of profitable and interesting ways
to invest in metals.

Chapter 10

Buying Metals Direct

· ·

In This Chapter

▶ Sizing up metals

▶ Checking out gold and silver bullion

▶ Finding dealers and resources

· ·

Most folks know something about gold and silver and can figure out to some extent what bullion coins and bars are. Although there are a wide variety of choices even in this apparently narrow area of the precious metals world, it actually won't be that difficult to choose. Buying and selling can be as easy as a phone call or a visit to a Web site. In this chapter, I explore the various ways you can buy metals directly.

Weighty Matters

The most common measure of weight that you'll come across is the ounce but we need to understand what kind of ounce. There are troy ounces and avoirdupois. My goodness . . . who thought that up? Probably the same guy who called cave exploring *spelunking*. Anyway, precious metals are measured in troy ounces.

Regular weight that you and I and the guy at the deli counter are familiar with is the avoirdupois. *Avoirdupois* (why don't they change that to *regular weight*?) is 16 ounces to a pound while that same 16 ounces is 14.58 in troy ounces. The troy ounce is roughly 10% heavier than the avoirdupois equivalent.

Bullion coins and bars are those in which you pay every cost above the actual metallic content. Gold may be $600 an ounce but that same ounce as a bullion gold coin or bar may be a few bucks more (due to fabrication costs). Bullion coins and bars are about as close as you can conveniently get to tracking 100% of the price of that metal, be it gold, silver, platinum, or palladium. Bullion coins and bars are generally bought purely because of the metal content and little else.

The Case for Physical Ownership

Physical ownership means having the actual asset in your possession. This contrasts with paper assets where most investors have greater familiarity. Examples of paper assets are stocks, bonds, and mutual funds. In the world of bullion there is paper in the form of silver (or other precious metal) certificates and pooled account programs.

In a paper asset, you have a claim on an asset via that piece of paper. The paper derives its value from the value of the underlying asset. For example, if you owned 100 shares of Gobbledegook Corporation, you own a small sliver of that company and the value of your shares rises and falls with the fortunes (or lack of fortune) of the company involved. If the public goes ga-ga over Gobbledegook (it happens more often than you think), then your shares will go up. If the company has problems or goes bankrupt, your shares will end up no better than bits of paper. Paper assets have potential problems quite different from owning the physical asset. For many investors, having the physical in your possession is more desirable.

Let's make the paper versus physical comparison more relevant since having, say, silver coins and having shares in a stock (even more specifically a silver mining stock) can be an example of comparing apples to oranges. Whenever investors are debating the merits of physical ownership, they are most likely talking about physical metal and *paper representing the physical metal.*

Since the 1970s, ownership in precious metals has taken various forms. The most obvious is the subject of this chapter which is physical ownership. In addition, there are competing forms of ownership through paper. Some examples are

- Silver certificates and warehouse receipts (can be allocated or unallocated)
- Perth Mint Bullion Certificates (unallocated)
- Perth Mint Bullion Certificates (allocated)
- Pooled Accounts (unallocated)

Note the terms allocated and unallocated. Allocated to metal means that it is segregated from other assets held by that company. It usually has its own serial number and you are charged more for this type of arrangement. Unallocated means that you have an ownership share along with others in a large pooled amount of metal that is not segregated and is not separately numbered. It's like when you go to a baseball game and you see the difference between prestigious box seats and the bleacher seats out in right field. Well . . . I hope you got that. Anyway, the unallocated usually has lower transaction and storage charges. The safer of the two is allocated.

For those who want to consider these venues then at least make sure that the companies you deal with have a long and unblemished record. A good example is the Perth Mint (based in Australia; visit its Web site `www.perth mint.com.au`). Some other places to consider are Sunshine Mint, Monex, and Kitco, Inc.

What could go wrong with paper assets

When I talk about what could go wrong with paper assets, I emphasize the point that there is a clear distinction between physical metal and paper representing the physical metal. You'll run into both. Because metals-related paper investment vehicles come in many forms, I think that it is helpful to summarize what could go wrong with paper assets versus the physical metal.

The most obvious potential pitfall with paper assets is the risk of default. The second potential pitfall is fraud. When you buy physical silver, for example, and you have it delivered . . . you have the physical silver. Period! With paper silver we need to keep in mind that the paper is a promise to deliver the underlying asset or to otherwise have a claim to that asset or to beneficial ownership, be it silver, gold, or other precious metal. For potential pitfalls of the various categories of paper assets, please review the accompanying Table 10-1 below.

Table 10-1	Paper Assets and Potential Risks	
Type of paper asset	*Potential risks*	*Get more details in*
Certificate for pooled account	Default, fraud	This chapter
Silver (or gold) certificate	Default, fraud	This chapter
Unallocated account	Default, fraud	This chapter
Warehouse receipts	Default, fraud	This chapter
Stocks	Company bankruptcy, financial difficulties may cause falling stock price, and so on	Chapter 12
ETFs	Market risk based on what happens to mining stocks in portfolio	For stocks, Chapter 12; for ETFs, Chapter 13

(continued)

Table 10-1 (continued)		
Type of paper asset	*Potential risks*	*Get more details in*
Mutual funds	Same as stocks except choices made by the investment firm; fees may erode returns	Chapter 13
Futures contract	Volatility, default	Chapter 14
Options	Could expire worthless; if you're wrong could drop in value drastically	Chapter 15

Bullion versus numismatics

Let's get physical; how does bullion stack up versus numismatic? The prime benefit of buying bullion is that you get 100% correlation with the asset itself. If you have silver bullion and silver goes up 20% or down 10% then your silver bullion will perform accordingly. It is a pure play on the asset.

Numismatics involve far more than just the metal content. The metal content, even if it was a precious metal, may still be a secondary or even minor factor in the value. A good example is some uncirculated gold double eagles. You could easily see a coin valued at $50,000 but the intrinsic metal value portion of the coin could be far lower. That coin may only have an ounce of gold and if the market price of gold is $650 then the intrinsic metal value would be only $650. What would account for the remaining $49,350 of value in that double eagle? The other factors of course, such as scarcity, condition, popularity, and so on. Chapter 11 goes into greater detail about numismatic coins.

The downside to numismatic coins can be the cost and the potential for problems in areas such as grading and dealer markups. Numismatic coins have a wholesale and a retail price that could have a wide margin that could be 30% or more (the premium over dealer cost). Bullion coins have a lower premium that could range from 2% to 10%. Bullion bars could have an even lower premium. Of the two, the coins have a higher premium because they aren't just lumps of metal; they are minted coins so the cost of fabrication is a factor.

When you are buying bullion coins, you pay a small premium over the metal content cost and it could vary depending on the dealer. Also keep in mind that when the time comes to sell your coins, you will get the price of the metal and some premium as well. Of course you will receive a lower premium than you paid but at least you get something extra. This is why bullion coins (and especially numismatic coins) should be held for the long term so that there is enough appreciation to hopefully more than offset the transaction costs you incur at the time of purchase.

Other metals

Owning physical metals is usually a reference for gold and silver. Having gold and silver is a practice dating back to the dawn of civilization but . . . how about other metals? Owning other metals directly in bullion form is not a large market at all. After gold and silver, the next one is platinum but it is a distant third. Recently, palladium has become available as a bullion investment.

Beyond those four precious metals, there's really nothing else in a convenient venue for physical ownership. What else could there be? Uranium might as well be kryptonite. Base metals are principally available as numismatic items and even then the base metal content is not the primary catalyst for investors. Is there such a thing as bullion base metals? Naah . . . not really. Nickel, for example, is measured in tons. Unless you have a really big yard, nickel and other base metals in bullion form are not feasible and out of the question. Even aluminum is out of the question . . . unless you're keeping some in your kitchen cabinet for wrapping food.

The risks of owning physical

The biggest concern of owning the physical outright is safeguarding it. Keeping the physical entails two risks:

- ✔ **Someone could steal it.** This is the most likely risk.

- ✔ **The government could confiscate it.** This may seem unlikely or very improbable but it's not impossible. History has taught us that government encroachment has happened before and it could happen again. Heck . . . if I told you in 2004 that local governments would have the power to take the home of a law-abiding, tax-paying citizen and that person could have no recourse (the eminent domain issue, authorized and expanded by the Supreme Court in 2005) you would have thought I was crazy.

In safeguarding your precious metals, how you secure it is also important. If you have gold and silver stored in a bank safe deposit box, the government can confiscate under certain conditions. Many folks prefer putting it in a secure place at home or their business. Use your discretion.

If keeping precious metals at home is a little unsettling for you then consider an alternative that many precious metals investors have used in recent years: James Turk's Gold Money program at www.goldmoney.com. The metals are allocated and insured.

Forms of Gold Physical Bullion

The first area of bullion to consider is with the ancient metal of kings: gold. With its beauty, reputation, and long history as a store of value and safe haven investment, it is a great experience to hold some in your hands. Fortunately, there are plenty of ways to do it, starting right here in the U.S.

One of the first things you learn about is karats. Also known as carats but definitely not known as carrots. The *karat* is a measure of fineness. A karat in gold is equal to ¼₄ part of pure gold (in an alloy). Here is a rundown of how many karats you need to be aware of:

- **24 karat gold:** Equals pure gold (100% gold).
- **22 karat gold:** ²²⁄₂₄ gold and ²⁄₂₄ other (such as copper). A bullion coin or bar that is 22 karat gold is roughly 92% pure gold.
- **18 karat gold:** ¹⁸⁄₂₄ gold (the other ⁶⁄₂₄ is copper or other metal). Basically 75% of the item is pure gold.
- **14 karat gold:** ¹⁴⁄₂₄ of the item is gold. In other words, 58.3% of it is pure gold.

One hundred percent pure gold is too malleable (soft) for regular use and it needs to be hardened by alloying it (mixing or combining it) with another metal such as copper or silver (or other metal). Twenty-four karat jewelry, for example, needs to be handled with much greater care than twenty-two karat (or lower) jewelry. This is why I try to get my wife to buy zero karat jewelry because . . . you know . . . it's a heck of a lot more durable . . . really. But I haven't succeeded in presenting my case. Oh well . . .

American eagle gold bullion coins

For gold investors, this coin is the first bullion vehicle to look into. In 1986 the United States Mint first starting minting the Eagle coin series (22 karats). The gold eagle caught on very fast and it's the top selling gold eagle in the country today. They are beautiful coins. The coin's design echoes the original design of the $20 gold eagle done by Augustus Saint-Guadens as commissioned by then-President Teddy Roosevelt.

The authenticity, content, weight, and metallic purity of gold eagles are guaranteed by the United States government, which makes them an acceptable investment not only to American investors but in international markets as well.

The advantages of gold eagles are numerous. Many financial advisors see them as a good way to diversify the typical portfolio. Gold is well known as an inflation hedge and gold eagles are a convenient way to add this benefit to long-term investment strategies. Gold eagles are very popular and the market for them is very large so buying and selling them is not a problem. Buying or selling them is usually private and nonreportable when done in small quantities.

Gold eagles are typically sold as one-ounce bullion coins. However, you can also get them in half-ounce, quarter-ounce, and $\frac{1}{10}$ ounce denominations. This makes gold affordable for just about any portfolio.

The American Buffalo Gold Bullion Coin

As gold rose in value in recent years, the interest in gold bullion picked up. Since the gold eagle was hitting record sales levels, why not other gold coins? It was at that point that the newest gold bullion kid on the block showed up: the 24-karat American Buffalo Gold Bullion Coin.

The face value of the coin is $50 and for the first time ever investors can get one troy ounce of pure gold in bullion coin form. These coins were manufactured by the U.S. Mint at West Point and they come in special protective holders. (Remember that pure gold is soft.)

The Krugerrand

Issued in 1970 by South Africa, the Krugerrand was the world's first gold bullion coin. Since their issuance, nearly 55 million coins have been bought and sold across the globe. The 22-karat coins are available in 1 oz., $\frac{1}{2}$ oz., $\frac{1}{4}$ oz., and $\frac{1}{10}$ oz. denominations. Because they have a huge market and are highly liquid, buying and selling them are easy.

The Canadian Maple Leaf

Issued by the Royal Canadian Mint, the Maple Leaf is one of the world's most popular gold bullion coins. Guaranteed by the Government of Canada for its authenticity and metallic content, the Maple Leaf has a purity extremely close to 24 karats: It's 99.99% gold, making it the world's purest gold coin. In addition to its reputation and quality, the Maple Leaf is also very liquid and accepted internationally.

Other gold bullion

Gold bars are popular and safe choices for investors and they have relatively lower premiums than minted coins. It's a good idea to buy them from sources that get these bars manufactured by Pamp Suisse and Credit Suisse. The gold bullion bar available sizes are

- 400 oz. bullion bars
- 100 oz. bullion bars
- Kilo bullion bar (32.15 oz.)
- 10 oz. bullion bars

Silver Physical Bullion

I like silver and I am very bullish on silver. I think that long term it is a good part of virtually any portfolio. Silver bullion is the best and easiest way to start participating. I give you several ways to get involved in silver in the following sections.

American Eagle Silver Bullion Coins

The U.S. Mint started minting these beautiful coins in 1986. With nearly 135 million coins sold since then, the American Eagle Silver Bullion Coins have become the world's best-selling silver coins. The design of these 1-oz. coins (.999 silver content) was inspired by Adolph A. Weinman, the designer of the 1916 Walking Liberty half dollar.

A primary advantage of the silver eagle is that it is very liquid so buying and selling are easy and convenient. A primary disadvantage is that the premium can run as high as $2 over the metal content value (depending on the dealer).

One-ounce rounds

These bullion coins weigh one ounce and they are .999 pure silver. They're like the silver eagle, but they have lower premiums. Private mints create them and the premium is typically 40¢ to 50¢ above the metal content value. Silver rounds aren't as well known as silver eagles, but they make good choices for investors who want to get more silver bang for the buck.

Junk silver bags

This may sound like an odd way to invest in silver but it is actually a great consideration. A junk silver bag has $1,000 of face-value silver coins issued in 1964 or earlier. The reason it has that less-than-appealing name is because the coins in the bag are in poor condition and you won't find any coin that is rare or has numismatic value. The coins are generally worn down from usage and would not be considered at any numismatic grade above good. For more information about grading and numismatic coins, see Chapter 11.

The $1,000 bag weighs about 55 pounds and it contains about 715 oz. of pure silver. What kinds of coins are in the bag? If half dollars were used there would be 2,000 coins. If the coins were quarters then there would be 4,000 coins. Dimes? Then there would be 10,000 coins. Because the bag of coins is bought as a bullion investment, the price of the bag would move in correlation with the price of silver. Many firms can sell you half bags that have $500 of face-value silver coins to make it easier for storage and handling.

Even though these old coins are generally priced as bullion, they are coins that are no longer minted. That tells you that there is a finite supply so prices for these bags could rise further if demand increases.

The difference between the buy and sell is fairly constant and there are investors who buy bags when premiums are cheap or negative and trade them when premiums increase substantially. Keep in mind that you're required by law to report your sale of $1,000 face 90% bags on IRS Form 1099B. Smaller quantities of 90% are not reportable.

The $1,000 bag of silver dollars

This interesting play in silver is not exactly a bullion investment but it is not a rare coin investment either. Circulated silver dollars struck between 1878 and 1935 carry higher premiums, yet they are a prime source of legal tender silver coins because they are so recognizable. These early dollars are divided into two price categories. The more expensive Morgan dollars struck between 1878 and 1904 and the Peace dollars struck between 1921 and 1935.

The Peace dollar being the less expensive is quoted by the bag. Such coins are always in average condition and grade VG (Very Good). A Morgan dollar bag and better quality coins are available at slightly higher prices so it pays to ask questions before placing an order. This option is popular because silver dollar bags have represented real, portable wealth for more than 100 years.

The 40% silver bag

The last silver coin the United States made for general circulation was the 40% silver clad 50¢ struck from 1965 through 1969. This is another popular way to own silver bullion in legal tender form. Like circulated 90% coins these $1,000 face bags are traded primarily for content. Because they are 40% pure a bag contains substantially less silver (296 troy oz.) which is reflected in a lower selling price.

The buyer has a number of advantages. First these are real U.S. coins and therefore are legal tender of our nation. In an emergency this could be significant. Second, the bag has a high face value ($1,000) which limits the money anyone could lose should silver move lower. This is easily seen in a down market because premiums almost always move higher. And finally, unlike 90% silver bags, you are not required to fill out I.R.S. Form 1099B on 40% bags when you sell.

Silver bars and ingots

In terms of getting the most silver metal content against the purchase price (in other words, paying the least amount of premium over metal content), bullion bars are the way to go. Ingots are really just small bars that are imprinted with designs to make the bar attractive and/or collectible. Ingots are more appropriate for collectors rather than for investors. In this segment we'll concentrate on bars (uh . . . bullion bars not adult drinking places).

The common popular brands that most investors will come across are Engelhard (www.engelhard.com) and Johnson-Matthey (www.matthey.com). There are more obscure sources but these two have a long-time reputation for their standards such as metallic quality and authenticity.

1,000 oz. silver bars

These bars tip the scales at about 68 pounds. They are typically used to settle the delivery obligations of futures contracts at NYMEX (the New York Mercantile Exchange; more about futures contracts in Chapter 14). A typical futures contract is tied to 5,000 oz. of silver so five of these bars will cover the physical requirements for delivery. The bottom line is that this size is not practical for average-sized or small transactions typically done by investors. Besides, why get sued by the delivery guy for his backache?

100 oz. silver bars

Another popular way to own bullion silver is the 100 troy oz. bar which is .999 fine. At a tenth of the size, it is also a tenth of the weight: 6.8 pounds. This is the most typical weight for large retail transactions among investors.

10 oz. silver bars

These are popular and the most common for small investors. They come with a slightly higher relative cost but as an absolute transaction, fit the small investor's budget.

There are other bar sizes (such as odd-weight retail bars) but these are the best sizes due to their acceptability in the marketplace and the high degree of liquidity (being able to convert to cash).

Platinum bullion

Platinum is the third most well-known precious metal, yet (according to industry reports) it is more rare than either gold or silver. Platinum is 15 times more scarce than gold! What better precious metal to consider adding to your portfolio? One quality that platinum shares with silver is that there are a variety of industrial uses for it. Its rarity and utility give it a bright future. There are several ways to invest in platinum bullion:

- ✔ Platinum American Eagle Bullion Coin
- ✔ Platinum Bullion Bars
- ✔ Platinum Canadian Maple Leaf Bullion Coin

Palladium bullion

Palladium used to be too obscure to invest in. It was the precious metals' equivalent of the one Marx brother whom no one knew (Grouch, Chico, Harpo, and . . . ah . . . Gummo?!), but it has gained respectability and quickly became, after silver and platinum, the "other white metal." You won't find much in the way of palladium coinage in general but there is palladium bullion that you can purchase and invest in. As palladium becomes more attractive in the ongoing precious metals bull market that started early in this decade, it can become a part of your portfolio. Here are ways to invest in palladium bullion:

- ✔ Palladium Bullion Bars (manufactured by companies such as Pamp Suisse and Credit Suisse)
- ✔ Palladium Canadian Maple Leaf Bullion Coin
- ✔ Palladium Chinese Panda Bullion Coin

For more details about palladium, see Chapter 7.

Bullion's costs and fees

When you are buying bullion coins or bars, you will come across "bid" and "ask" prices. The difference between the bid and ask amounts is the spread. Generally, coins and bars that are actively bought and sold have a relatively small spread. Coins and bars that are in a slow or thinly traded market tend to have wider spreads. This you need to know; when you are buying, you pay the ask price. When you are selling, you will receive the bid price. The bid and ask prices vary greatly from dealer to dealer.

In addition, find out about fees that cover costs such as shipping, storage, and any transaction fees. The size/quantity of the order will matter. The premium per ounce on a 1,000-oz. bar order will be relatively lower than the premium on a 100-oz. bar order, which in turn will be relatively lower than for the 10-oz. bar order.

Even in the realm of precious metals, the advice from every consumer guide in the history of humanity says "shop around for the best price." Hey . . . the best advice isn't always the most original!

So, what kind of silver is best?

Some of this is tied to your preference. Those who are purists like the bars and one ounce rounds. Those who want a more popular, more liquid, and easier bullion transaction would opt for the U.S. Eagle coins. Your budget will also dictate the transaction. Some buy a large, lump sum now. Still others will allocate a portion of their monthly budget and do some dollar-cost averaging.

I think that serious investors would consider a portion into bullion as a part of their precious metals foundation and then allocate a portion of their investable funds among some of the paper assets (such as mining stocks and ETFs). Over the long haul, accumulating your investing among the different classes of physical and paper assets will prove rewarding as the precious metals bull market unfolds in the coming months and years.

Gold and silver commemorative coins

I could have easily put the topic of commemorative coins in the numismatic chapter, but I thought it would have a better fit here. Why? Commemorative coins were not issued for circulation and the modern commemoratives had greater value due to their metal content. But these coins can indeed be collectible. These coins could easily become a beer commercial debate. (Tastes great! Less filling!)

The real answer is that commemorative coins have something to offer both the collector and the bullion investor. There are two types of commemorative coins: early commemoratives issued 1892–1954 and modern commemoratives issued from 1982 to the present. As the folks point out, the early commemorative coins are highly collectible (they have both numismatic and investment value), while the vast majority of modern issues are not. Some modern commemoratives have taken off such as the 1997-W gold Jackie Robinson $5 half eagle. By mid-July 2007, that coin was trading in the $5,000 to $6,000 range. The vast majority of modern commemorative coins do not fetch any substantial value beyond their metal content. For serious investors, the early issues are where the numismatic and investment action is.

The public is offered newly issued commemoratives that are very overpriced. Frequently the price is ten times (or more) the value of the actual bullion content if it is marketed with a recently deceased celebrity or famous event. It may be touted as a collectible but it is either way overpriced or simply a rip-off. The bottom line is that if you see it being marketed in expensive venues such as television, then the items sold will be too expensive.

Among the more popular commemoratives are the 1892 Colombian Exposition half-dollar (the first commemorative coin authorized by Congress) and the half-dollar commemorating George Washington-Carver (issued during the years 1951–1954). During 1892–1954 the U.S. Mint produced commemorative coins for 53 different events and honored individuals on a total of 157 different gold and silver coins.

The same sources (listed throughout this chapter) that sell bullion coins usually sell commemorative coins. For more information on quality commemoratives you can turn to the sources listed in Chapter 11 (on numismatic coins) as well as the U.S. Mint (www.usmint.gov) and the Society for United States Commemorative Coins (www.suscconline.org).

Bullion Dealers and Resources

Fortunately, finding dealers and shopping around is easier than ever before. Now you can search for a coin shop near you, online. The following are some dealers with Web sites to help you do your shopping:

- ✔ Bullion Direct (www.bulliondirect.com; 512-462-2646)
- ✔ Northwest Territorial Mint (www.nwtmintbullion.com; 800-344-6468)
- ✔ American Gold Exchange (www.amergold.com; 800-613-9323)
- ✔ Investment Rarities (www.investmentrarities.com; 800-328-1860)

  ✔ To find a local coin shop, you can search by name, by state, or by zip code at www.coininfo.com.

  ✔ For more on what kinds of silver to buy, and where to get it, see www.find-your-local-coin-shop.com.

To get consumer reports on how to avoid scams in precious metals or buying tips on silver (and other) bullion, go the Silver Stock Report (www.silverstockreport.com). It is a newsletter edited by silver analyst Jason Hommel, a silver analyst and editor of the Silver Stock Report. He has done extensive research on the topic of paper versus physical and shares his research through reports available at the Web site. His personal preference is silver bought as bullion bars, rounds, and junk silver bags. Hommel is a staunch proponent of owning (and taking delivery of) physical silver. He regularly publishes consumer reports on buying physical silver.

Putting Precious Metals in Your IRA

If you think that paper assets (stocks, bonds, mutual funds, and so on) are the only things that can sit inside an IRA (Roth or traditional), think again. Actual physical bullion assets (gold, silver, and/or platinum) can now be part of your retirement strategies under the tax-advantaged umbrella of your IRA.

Precious metals allowed for IRAs

The following coins are examples of the types of coins that are allowed to be placed in an IRA:

  ✔ American gold, silver, and platinum Eagle Bullion Coins

  ✔ Canadian gold, silver, and platinum Maple Leaf Bullion Coins

  ✔ Gold, silver, platinum, and palladium bars and rounds manufactured by a NYMEX or COMEX approved refiner/assayer and meeting minimum fineness requirements. Can be 1,000 oz. or 1,000 oz. bullion bars.

  ✔ Other coins that *specifically* meet the IRS requirements for inclusion.

Here are some types of precious metals investments that are *not* allowed to be placed in an IRA:

  ✔ Krugerrands (gold bullion coins minted by South Africa)

  ✔ Old U.S. Gold numismatic or collectible coins

  ✔ U.S. 90% silver coins (1964 and before)

The IRS laws covering precious metals IRAs can change from time to time so check on the latest information at www.irs.gov. More on taxes in Chapter 20.

Establishing an account

Get the full details through organizations that have met IRS regulations which govern retirement accounts. As of June 2007, the following organizations have been approved for precious metals IRAs:

Sterling Trust Company

PO Box 2526 Waco, TX 76702

7901 Fish Pond Rd., Waco, TX 76710

(254) 751-1505

(800) 955-3434

(254) 751-0872

www.sterling-trust.com

American Church Trust Company

14615 Benfer Road

Houston, TX 77069

(800) 228-8825

Fax (281) 444-9797

www.churchtrust.com

The first organization you should contact is the bullion/coin dealer to find out which of the above companies it works with. Once you know that, contact them directly and find out about the paperwork and fees involved.

The actual ordering of the approved coins and bars will occur through the dealers and they will work with the IRA firms for logistics such as where the bullion is to be shipped for approved storage. The approved storage is an important part of the arrangement because without it, you couldn't have the precious metals IRA at all.

Chapter 11

Purchasing Numismatic Coins

*N*umismatics. Who thought up that name? Why not call coin collecting . . . uh . . . coin collecting? And then there's stamp collecting: philately (probably from the same re-naming genius). Why stop there? Why not call wine tasting *grapeology* or skydiving *splat-o-lympics*? Anyway, coin collecting . . . uh . . . numismatics . . . is a great and interesting way to get into precious metals. Don't confuse it with bullion coins and bars (as covered in Chapter 10) since the metal content is only part of the reason that you get into numismatic coins.

In 2002, a 1933 gold double eagle coin sold for $7.5 million. Whew . . . now that's not chump change. But it does attest to the clout of investment-grade numismatic coins. Rare, high-quality coins can command prices of five, six, or even seven figures so it certainly pays to consider the power of coins in your portfolio, especially in an era of growing demand and inflation.

Numismatic coins offer some of the same advantages as other investment vehicles such as scarcity, liquidity, and appreciation. But another dimension comes into play; it can be an interesting and absorbing hobby. For years I was an active numismatist who was initially drawn by the opportunity for long-term appreciation but came to see the other appealing features of the world of coin collecting.

The Basics of Numismatics Coins

Numismatic coins are coins that have achieved value due to their rarity. As you may recall, bullion coins are primarily acquired for their metal content. For numismatic coins you need to consider (or be aware of) the following:

- **Metal content:** This is first only because this is a major consideration in the book. The major types of precious metals are silver and gold. There are also other metals as well (base metals such as copper and nickel) but this is not a factor in the value as is the case with precious metals.

- **Rarity:** The fewer there are of a particular coin, the greater the potential value. Many old coins are valuable because of their rarity.

- **Grade (or condition):** The better the condition, the higher the value. The grade is a crucial factor in the coin's value (more on this later in this chapter).

- **Age:** This is a relatively minor issue but it is worth listing. All things being equal, a 100-year-old coin has greater value than a one-year-old coin.

- **Popularity:** Some coin series are more popular than others. The coin's popularity may be attributed to its beauty or historical significance.

- **Mint mark:** Coins were minted at a variety of minting facilities throughout U.S. history. A coin in the same year but from a different mint could be more scarce, hence more valuable.

As you can see, numismatics can be a little bit more complicated than just the age and/or metal content. Of course if you have an old coin made of a precious metal such as gold or silver that is in excellent condition and is rare and popular then you have a winner!

Profitable coin investing

If you are going to be successful in coin investing (certainly financially successful) there are some golden rules that will enhance your efforts:

- **Stick to precious metals:** Because of other factors covered in this book (such as inflation), it will enhance your long-term profitability to stick to gold and silver coins due to the metals' appeal for a variety of reasons. As contemporary coinage becomes more debased (the government is using cheaper metals to keep coin mintage costs low) that means that more valuable coins will keep rising in price.

- **Specialize:** It's hard to keep track of all the coin series. It is advisable to stick to a single popular series (certainly in the beginning anyway) such as Mercury dimes or Morgan dollars. Get to know the key dates and grades.

✔ **Quality:** Buy the higher grades since they will fetch a higher price. Investors will generally look at the uncirculated and proof grades first (see grading explanations below). Depending on the year and mint, an uncirculated Morgan silver dollar, for example, could easily be worth thousands of dollars more than the same coin in Good or Fine condition.

Making the grade

Grading is a reference to the coin's physical condition. The grading system, referred to as the Sheldon Scale (see Table 11-1; named after William Sheldon who standardized coin grading in 1948) is an industry standard that helps dealers, collectors, and investors find an easier way to determine the coin's condition. The Sheldon Scale works on a numeric system ranging from 1 to 70, with 70 being the highest and most flawless level.

Table 11-1		Sheldon Scale Rundown of Grades
Level	*Grade*	*Comments*
AG - 3	About Good	Lowest grade. You can barely make out the features on the coin. This is fine if you are seeking coins for their metal content but the worst (and cheapest) choice for numismatic investors.
G - 4	Good	This is not good in the true sense. It is a notch above the worst. This is a poor condition.
VG - 8	Very Good	
F - 12	Fine	The fine grades are still low-grade. Much better condition than the good grades but not investment-grade.
VF - 20	Very Fine	
VF - 30	Choice Very Fine	
EF - 40	Extra Fine	
EF - 45	Choice Extra Fine	Choice Extra Fine is okay if you're talking steak, but in the world of coins, it is not investment grade; it's fine if you are just a collector.
AU - 50	About Uncirculated	Now you're talking. Uncirculated means that the coins are in excellent condition. All the features are strong with very little wear and some nicks.
AU - 55	Choice About Uncirculated	

(continued)

Table 11-1 *(continued)*

Level	Grade	Comments
AU - 58	Very Choice About Uncirculated	
MS - 60 to 70	Uncirculated or Mint State	This should be the lowest level for investors seeking coins with high desirability. The better the grade, the higher the price at sale time.
MS - 70	Proof	This is the top grade. The coin has a mirror-like look and is in superb condition with no blemishes, nicks, or other signs of wear or contact. See below.

The grades of MS - 1 to MS - 59 could really be called the collectable grades. For collectors who are simply seeking to add to or complete their coin collections, these grades are okay. Because the grades are low, the prices are also generally low so coins at these levels are affordable.

Those seeking coins with the best potential for investment gain need to concern themselves with the higher grades of MS - 60 to MS - 70. To get a truly good idea about these investment grades, it's best to get it from the source, of course. You can find full descriptions straight from the Fifth Edition of *Official ANA Grading Standards for United States Coins* published by the American Numismatic Association.

Information sources

As with most things in life, the more information you have, the better off you're going to be. The top sources of information on numismatic coins are

- Krause publications. Krause (www.krause.com) publishes *World Coin News*, *Numismatic News* and *Coins Magazine*.
- Coin World (www.coinworldonline.com)
- Coin Resource (www.coinresource.com)
- About.com (www.coins.about.com)
- Daily Numismatic & Gold Investment News (www.cointoday.com)
- All Coins Numismatic Resources (www.allcoins.org)
- Numisma-Link (www.numismalink.com)

Collectible Coins

Once you get a good understanding of the coin market and important features such as the grade (quality) and mintage (quantity), you can choose what coin series you want to get involved with. In this segment, I won't cover every series that is out there since there are lots of coins to deal with. It is important to stick to those series that have a large, popular market since it will make it much easier to buy and sell when the time comes.

Gold coins

Although the U.S. Mint (`www.usmint.gov`) has minted gold bullion coins in recent years, the last numismatic gold coins were issued in the early 1930s. One of the most popular gold coins is the $20 Saint-Gaudens coin issued 1907–1933. It is still a very popular coin for collectors and investors and it serves as the quintessential gold coin. Here are the most popular numismatic gold coins:

- The $20 Saint Gaudens double eagle
- Liberty head $1 gold
- Liberty capped bust 2½ dollar gold
- Indian 2½ dollar gold
- Indian $5 gold
- Indian $10 gold

By the way, some of the coins are referred to by the designer's name. The $20 gold double eagle coin was designed by Augustus Saint-Gaudens. Later in this chapter you see a reference to coins such as the Barber halves. Before you look for the razor and shaving cream in the design, keep in mind that it is another reference to a coin designer, Charles E. Barber. Barber designed many of the coins in the late 1800s.

Silver coins

Silver coins were more prevalent than gold coins and they were in common circulation until 1964. From the early days of our country's history to 1964, we have had silver dimes, quarters, halves and dollar coins. These coins were 90% silver. An exception was Kennedy halves that still contained 40% silver during 1965–1970. After that, all common coins in circulation were made from base metals.

The most common silver coin series are

- ✔ Morgan Silver Dollars (1878–1921)
- ✔ Peace Silver Dollars (1921–1935)
- ✔ Trade dollars (1873–1885)
- ✔ Liberty seated dollars (1836–1873)
- ✔ Early silver dollars (1794–1804)
- ✔ Franklin Halves (1948–1963)
- ✔ Walking liberty halves (1916–1947)
- ✔ Barber halves (1892–1915)
- ✔ Liberty seated halves (1839–1891)
- ✔ Liberty bust halves (1807–1838)
- ✔ Early halves (1794–1807)
- ✔ Washington silver quarters (1932–1964)
- ✔ Standing liberty quarters (1916–1930)
- ✔ Barber quarters (1892–1916)
- ✔ Liberty seated quarters (1838–1892)
- ✔ Draped bust quarters (1796–1838)
- ✔ Roosevelt silver dimes (1946–1964)
- ✔ Mercury dimes (1916–1945)
- ✔ Barber dimes (1892–1916)
- ✔ Liberty seated dimes (1837–1891)
- ✔ Liberty bust dimes (1796–1837)

To get much more detail on these and other coin series, consult the various coin resources and publications in this chapter.

Other coins

Beyond gold and silver (there are no platinum numismatic coins, only platinum bullion coins as in Chapter 10), there is little else in the way of precious metals for collectible coinage. However, you will find plenty of base metals used. Copper, zinc, and nickel are common in coinage and there are plenty of choices here for collecting and investing. Here are the most common collectible series with base metals:

- The Lincoln wheat cents (1909–1958)
- Jefferson nickels (1938–present)
- Buffalo nickels (1913–1938)
- Liberty nickels (1883–1913)
- Shield nickels (1866–1883)
- Indian cents (1859–1909)
- Flying eagle cents (1856–1858)
- Large cents (1793–1857)

Keep in mind that even coins with base metals have risen in value very nicely in recent years. The same things that have a positive impact on precious metals (such as money supply growth and price inflation, for example) have a similar impact on base metals. A good example is the Lincoln wheat cent.

The original Lincoln cent was 100% copper. During World War II, copper was so necessary for the war effort that Lincoln cents were made primarily with steel instead of copper temporarily in 1943. In recent years, the price of copper rose dramatically and in recent years copper is only a miniscule part of pennies today. The increase in the price of copper was just another reason for why the older Lincoln cents rose in value.

Still other coins: Commemoratives

One area of the coin world that I don't want to leave out is commemoratives. These are not coins in the normal sense; they are not issued for general circulation. Every coin that I have written about in this chapter could actually be used as regular money in that you can use these coins to buy stuff at the store but . . . *please* . . . don't do that! Anyway, commemoratives are just that; they commemorate a person, organization or event. Gold and silver commemoratives are a large market for collectors and investors. Some of the same rules apply (such as scarcity, metal content, and grade).

It is important to stay with commemoratives that have a substantial market so that buying and selling are easier. Stick with quality and with reputable dealers. For more information, check out the same Web sites and resources indicated in this chapter. Some sources for information on commemoratives are the Coin Alley (www.thecoinalley.com) and the U.S. Mint (www.usmint.gov).

Coin Services and Organizations

If you are serious about numismatics as a long-term investment vehicle, then it pays to join one or more of the organizations in this specialized area. As you will find out, numismatic coins are more complicated than bullion coins and you need to keep informed. The first place to start is the associations:

- American Numismatic Association (ANA; www.money.org)
- The American Numismatic Society (ANS; www.amnumsoc.org)
- Professional Numismatists Guild (PNG; www.pngdealers.com)

The associations provide a lot of information and guidance to collectors and investors. Member benefits usually include newsletters, books, and conferences. Meeting other members is easy and you can quickly get information from others on the particular coins that you are focusing on. Speak to experienced members and ask a lot of questions about grading, mintage, rarity, selling tips, and so on.

Information sharing

Besides the associations, the Internet has another venue for you to meet folks and exchange opinions and information on numismatics. Use your favorite search engine to find newsgroups. Some of the more active ones are Coin Forum (www.coinforum.org) and Coin Talk Forum (www.cointalk.org).

Grading services

Because grading can be such a touchy and delicate matter — a single level difference in grade could mean thousands of dollars — it is recommended that you consult a grading certification service. A reliable grading certification service is crucial whether you are buying or selling your coin. The service can not only certify the grade of the coin, but also authenticate it as well. The following are the major coin grading services in the industry:

- Numismatic Guaranty Corporation (NGC; www.ngccoin.com)
- Professional Coin Grading Service (www.pcgs.com)
- American Numismatic Association Coin Service (www.anacs.com)
- Independent Coin Grading Company (www.icgcoin.com)

Generally they will request that you fill out their form and send it in with the coin(s) and their fee. Depending on how busy they are, expect to receive your coin back with their certification in several weeks or longer. Here are some questions that you should consider asking right after "How much does it cost?" and "When will my coin collection be worth a million bucks?" . . .

> ✔ How many years have you been grading and authenticating coins?
>
> ✔ Does your firm offer a guarantee in your grading and authenticity service?
>
> ✔ What type of holders (containers) do you use to safeguard the coins?
>
> ✔ What grading standard do you use?
>
> ✔ What established dealer or collector groups use your services?
>
> ✔ Where do your firm's certified items trade (auctions and so on)?
>
> ✔ What's the capital of Bolivia?

Selling Your Coins

Understanding what you are buying is important but what better way to end the chapter than understanding selling your coins. After all, the whole point of buying collectable coins (at least from an investment perspective in this book) is to gain by selling it at a higher price later on. For maximum investment gain, it is best to hold your coins long-term to give them adequate time to appreciate in value. Coins are not really appropriate as tradeable vehicles.

Keep in mind that to achieve maximum gains with your coins, the issue isn't just when to sell (again, later is better than sooner). It is also important to understand whom you sell to.

Back to dealers

The most convenient way to cash in your coin investments is to sell to a coin dealer. The reputable ones are members of the long-standing organizations such as the Professional Numismatists Guild and you shouldn't have a problem making a fast and efficient sale. However, keep in mind that what you are seeking to get paid is not the same as what dealers are willing to pay.

I'm sure that after you poured over all the price listings in coin publications you have a preconceived idea of what your coins are worth. Keep in mind that you are probably looking for the retail price while the dealers are only looking to pay a wholesale price. Oh sure, you read the dealer's ad stating "We pay the highest prices in the industry" and it may even be 100% true but in reality the point is highest wholesale prices. Dealers are in business and they need to make a profit to stay in business. If you think that the value of your coin should be, say, $100 (that's what you saw in those coin publications) then the amount you will likely be offered will be in the $50 to $65 range. This is why you should consider your coins to be a long-term investment to give them enough time to appreciate. The difference will mean another 30% to 40% in the price to cover the difference between wholesale and retail.

Of course, if the coins you are seeking to sell were found in your attic or inherited, then the wholesale/retail price difference may not mean much to you since your purchase price was effectively zero.

Dealers will generally pay a wholesale rate for your coins. Therefore, if you are settling for dealers to make a fast and convenient sale, call several dealers to size up the offers.

To other investors

If you want to gain a higher price, then it may behoove you to skip past the middleman and go straight to the buyer . . . another investor. You may say "great! The heck with the dealer . . . I'll go direct to the buyer and get more money!" Here is where you discover the trade-off. If you want to realize more money from the sale of your coins, then you'll need to put in the time, effort, and diligence. Realizing a higher price entails more marketing and sales efforts on your part.

You can find coin investors in a variety of places and the Internet makes it easier to locate would-be buyers. There are numismatic clubs and chat rooms. There are markets for buyers and sellers of coins at Web sites such as www.allcoins.com. You can also go to coin shows and conferences (there are sites such as www.coinshows.com).

Selling directly to individuals will be easier if you get your coins certified. Investors are more apt to make a deal with you if a credible organization (see the grading services above) has certified the coin and its condition and authenticity.

eBay and other auctions

Online auctions are one of the most active areas of the Internet. Unless you've been spelunking (a weird word that means exploring caves) for the past dozen years, you've heard of eBay (www.ebay.com). It is, of course, the premier online auction site and lots of coins are bought and sold there regularly. There are lots of great information sources on how to buy and sell on eBay and it's the first place to look when you are considering selling your stuff online.

eBay is a horizontal auction site which means everyone sells all sorts of stuff to everyone else. If you can't sell your coins on eBay then consider finding a vertical auction site. This type of auction site means the buying and selling are in a narrow niche or specialty. There are auction sites that specialize in coins. You can find these sites by utilizing a major search engine (or you can go to a comprehensive site such as www.internetauctionlist.com).

Also keep in mind that there are dealers and auctions that do consignment sales. Consignment means that you sell your item through a sales agent but you still retain ownership of the item until the sale is made.

Here are some of the major coin auction Web sites:

- Bowers and Merena (www.bowersandmerena.com)
- Heritage Auction Galleries (www.coins.ha.com)
- Teletrade (www.teletrade.com)
- R.M. Smythe & Company (www.smytheonline.com)

Pricing information

Whether you are buying or selling, the price is everything. You shop around if you are going to buy. If you are going to sell, you need to verify with reliable sources regarding the market value of your coin(s). The following sources have pricing information that can help.

- Numismedia Fair Market Value Price Guide (www.numismedia.com)
- Coin Universe Price Guide (www.coinuniverse.com)
- Coin Net (www.coinnet.com)

A book that achieved renown as the bible for coin prices and values is the Red Book. The full title is *The Official Red Book: A Guide Book of United States Coins*. It is inexpensive and compact. Although it gets updated annually, that's enough. Coin values are not that volatile so an annual book is sufficient for most folks.

One final note . . .

Last but not least . . . check your pocket change everyday. You'd be surprised by what you can find that has greater value than you think. Many years ago when I was a kid, I went into a bank and gave them twenty bucks and asked for some rolls of halves. Being a coin collector and a hunter I always checked my change and even went to the bank to check their change, too! In those rolls of half-dollar coins, I was delighted to find over half of the coins were Kennedy 40% silver halves. Even today, you can still spot the occasional silver dime or wheat cent. It's like finding real money.

Chapter 12

Mining Stocks

A h yes . . . stocks. My old stomping grounds (shameless plug here for *Stock Investing For Dummies*)! Well . . . I still stomp here in my daily professional activities so you know why I'm misty-eyed about the topic. In this chapter, I get to give you a summary of some of my favorite (and most essential) stock investing points but also get to customize it for the world of precious (and base) metals. Some of your greatest profits (and risks) will be in mining stocks. For most investors, mining stocks are the easiest way to participate in precious metals investing and speculating.

Before we get to the information relevant for mining stocks, it will be useful for you to get some important points on stock investing.

Essential Stock Investing 101

I love stock investing. It is an appropriate venue for most investors and speculators. The act of buying and selling stock is easier and more widespread than ever before. What isn't easier is the act of choosing which stocks to initially buy. Successful stock investing isn't just understanding stocks . . . it's also understanding yourself and your goals. There are different kinds of investors just as there are different kinds of stocks. Before you get to the specific area of mining stocks, it is important to summarize what to look for in stocks in general.

Stock investing for all

The whole point of stocks is to choose one that has the greatest chance of having a rising share value. We all want to buy a $3 stock and sell it later on for $257 per share or better. But . . . how would you proceed to accomplish such a feat? What would make a stock rise so much?

Buying a stock is essentially buying a piece of a company and its potential for profit and growth. If the company does well, ultimately the price of the stock attached to it will rise. If the company does poorly, ultimately the stock will go down. Why would the stock go up (or down)? Always keep in mind that the real force that moves the price of the stock up or down is the marketplace. The marketplace is composed of buyers and sellers that are individuals and organizations. If there are more buyers (versus sellers) of a stock, then it will go up. If there are more sellers (than buyers) then the stock price will go down. The buying and selling goes on in stock exchanges such as the New York Stock Exchange (www.nyse.com) and places that are not technically exchanges such as Nasdaq (www.nasdaq.com). For you to conduct your stock transactions, you would need a stock brokerage account.

Why does anyone buy a stock? Of course, we can say the obvious, that you buy a stock because you think that the stock's price will go up and you can sell it later at a profit (capital gain). So the primary reason the public buys a stock is for gain. The second reason a stock is bought is for income (dividends). Ultimately, the stock and its performance will be based on the company's performance as measured by sales and profits. It seems like a pretty logical process: Find a company with strong prospects and potential success and the stock attached to it will rise and you will profit. Not so fast . . .

Sometimes you will see Company A doing well and being profitable yet the stock's price go down. Or you may see Company B doing poorly and losing money and yet . . . its stock rises. It makes you scratch your head. What gives? I can boil down to one word why anyone buys a stock: expectation. When you see the stock of a profitable company go down, it is obvious that there were more sellers than buyers . . . but why? The answer is that stocks don't go up and down because of how well (or how poorly) a company is doing presently; they go up or down because of what buyers or sellers *expect* will happen with that company in the near future.

Understand that the behavior of a stock in the short-term is irrational; the stock's price will behave in ways that are seemingly crazy and illogical. However, the performance of the company and the performance of the stock over the *longer term* do become a logical relationship. Long term, solid, profitable companies have rising stock prices while poorly run, unprofitable companies have declining stock prices. The lesson to investors is that trying to make short-term profits in their stock investing approach is a gamble (it is speculating) while making money is easier when you are patient and focused on the longer term. Sooner or later, the market figures it out. Good companies whose share prices have gone down in the short-term become a buying opportunity. Bad companies whose share prices have gone up in the short-term give you the opportunity to unload the shares.

The bottom line is to not worry about the short-term gyrations of the stock. Instead, focus your time and effort in finding companies that are strong, well-positioned in the right industries, and have solid fundamentals (good products or services, growing industry, rising sales, increasing profits, and so on). Let's focus on some of those fundamentals, such as . . .

A growing industry

Ever heard the expression a rising tide lifts all boats? Well . . . before you go buy a yacht, think over what that really means for your stock portfolio. Having tracked the stock market over the past few decades, I've seen it over and over again. I've seen the stocks of mediocre companies go up because they are in a strong and growing industry. And of course I've seen the reverse. Picking a stock doesn't happen in a vacuum. Understanding and being aware of that company's industry and the overall economic environment is critical to stock picking success. Sometimes I think that the industry and overall economy matter more than the company itself. I have avoided good companies because I expected their industry to experience tough times. I also had no problem buying shares of average or below-average companies that were in hot industries. You will also find out that figuring out the general direction of the industry is easier than for the individual company.

Become very proficient and knowledgeable on an industry that has good growth prospects. Real time (summer 2007), I am telling my clients and students the following . . .

> For conservative investors, consider consumer staples such as food, beverages, utilities, and established companies in energy and natural resources.

> For aggressive investors and speculators, consider energy, alternative energy (such as uranium, solar, and so on), and natural resources such as . . . precious metals! (You must have seen that one coming.)

Go do your research at a well-stocked business library or on the Internet at Web sites such as:

✔ Market Watch (www.marketwatch.com)

✔ Bloomberg (www.bloomberg.com)

A profitable company

What insightful advice: choose a company that's profitable! Well this point is not as obvious as it sounds. There have been many companies that were unprofitable whose stock rose dramatically. There's nothing wrong in buying the stock of a company that is losing money. But . . . if you do that, you're not investing, you're speculating. Investing means that you are putting your money into assets (stock, bonds, gold, and so on) that have provable value and that you understand how the investment works and that it is appropriate for you and your financial objectives.

Speculating is a different animal altogether. Speculating is a form of financial gambling. Now that may be too extreme a definition since speculating is a bit more complicated than tossing some dice at a casino table. However, the essence of the transaction is an educated guess. In speculating, the payoff could be big and fast. However, the potential loss could be equally big and fast. You should only speculate with a relatively small portion of your total finances. I personally enjoy speculating and I do a lot of it but I still keep the bulk of my money in more stable, predictable venues such as conservative stocks, cash, and so on.

Back to profitability. Profit is important. Actually, profit is the *most* important part of the company's total financial picture. It is profit that is the lifeblood of the company; profit is the engine of growth for innovation, jobs, and so on. I'll take it a step further; profit is the lifeblood of a successful economy. You show me an economy where profit is abolished and I'll show you an economy that either has or will collapse. It's true for the company as well. How do you find out about the company's profitability?

The profit is also expressed as net earnings, net gain, and net income. It is the end result in the company's income statement (sales less expenses equals net profit or net loss) which can be found in a number of places. The income statement (also called the P&L statement or profit-and-loss statement) can be found in the company's annual report or Web site. In addition, it is part of the financial statements that are made available at other venues such as at the Securities and Exchange Commission (go to www.SEC.gov and use the agency's EDGAR database to locate any public company's public filings) and many financial Web sites such as Market Watch, and so on.

As a general rule of thumb, investors should look at how profitable the company has been in the most recent three consecutive years. You want to see consistent profits and it is preferable that the profits are at least 10% (or more) higher than the year before.

In addition, you want to see rising sales in conjunction with those rising profits. Total sales should be consistently rising year-over-year (again, I like to use an easy benchmark like 10% or more).

Low debt? High assets?

The next thing to look at is the company's balance sheet. It is simply a snapshot at a point in time showing what the company owns and what it owes. It is expressed as assets minus liabilities equals shareholders' equity (or net assets or net worth). A company's financial statements are not that difficult to figure out since there is a lot of similarity to your financial picture. You personally have a budget and you need to track your income and outgo (just like a corporate income statement). You track what you own and what you owe (again, just like a corporate balance sheet). Look for a few things that are key to the company's financial health:

- ✔ **Is the company's net worth growing every year?** Each year should be at least 10% higher than the year before (preferably more).

- ✔ **Are assets increasing?** The more it owns, the better.

- ✔ **Are its debts stable, low and/or decreasing?** The less it owes, the better.

See . . . that wasn't so bad! Yes, there's always more to know but if you get the major numbers right, you'll be almost home free. Now, one more thing regarding the numbers . . .

Some ratios give you perspective

You need to take a close look at an intimate relationship. No . . . don't go peering into your neighbor's window . . . I mean the company's numbers! I am talking about ratios that help you analyze a company's finances to get a clear picture of the company's financial health. Here are some ratios that are critical for your analysis:

- ✔ **Price-to-earnings ratio (PE):** This is based on a per-share basis. You divide the price of a share price by the net earnings per share. The PE ratio establishes a connection between the share price and the company's bottom line. All things being equal, a relatively low PE ratio (under 15) is considered safer than a relatively high PE (over 30). Large, stable companies tend to have low PE ratios. Small, speculative companies tend to have very high PE ratios (some are over 100, sometimes way over!) which indicates greater risk. Companies with no PE ratio are the riskiest. (They have a share price but have no earnings or have losses, therefore no PE ratio.)

✔ **Debt-to-asset ratio (DA):** This ratio puts the company's total debt (or total liabilities) and total assets in perspective. For example, a company with total debt of $1 million and total assets of $2 million has a .50 DA ratio. In other words, the company has 50 cents of total debt for every dollar of assets. A DA ratio of .50 or less is desirable. A DA higher than 1.0 indicates a debt load that is too high and it could cause the company problems, especially if the economy is slowing down.

✔ **Comparative year ratios:** This is simple. You compare the important line items of one year to the prior year. If total sales this year are $15 million and last year they were $10 million, your sales are 50% higher. You do the same calculation with net earnings and with the important numbers on the balance sheet (such as the net equity).

There are excellent Web sites with information for beginners looking for explanations regarding ratios and other financial terms. Check out www.investorwords.com and the glossary at www.investopedia.com.

Getting more details . . .

That went by fast. There was so much I couldn't squeeze in but I do hope you continue reading up on stock investing; it can only help you. Here are some great places to turn to:

✔ **Value line:** Most business libraries have their excellent publications and you can check out the Web site of the same name.

✔ **Standard & Poor's:** Ditto.

✔ The **stock exchanges** have great resources, glossaries, and even some tutorials on stock investing. Go to

- www.nyse.com

- www.amex.com

- www.nasdaq.com

✔ **Yahoo! Finance** has lots of investing information and stock market data.

Mining stocks 101

The points above are indeed a mini-crash course in stock investing ("Paul, don't use the word *crash* in a stock investing chapter!") so it doesn't do the full justice to such a great topic so do catch up with other books. I mention the book *Stock Investing For Dummies* as a resource not because I wrote it but because it has all the important things that I think budding investors should know about. With that caveat and setting aside the basics of stocks in general, let's take a closer look the heart-and-soul of this chapter, the mining stocks.

It's all mine

Why do you buy the stock of a gold mining stock? How about a silver mining stock? Or any stock of any company involved primarily or exclusively in natural resources? It's what they have. It pays for you to find out about their properties and find out about their provable reserves. It can tell you if the stock's price is undervalued or overvalued.

Here's an example. I know of one silver mining company that was recently priced at $37 per share. Yet the company has in provable reserves over one billion ounces of silver in its properties. With silver at $13 an ounce, it means that its total silver reserves are valued at $13 billion. When you divide the company's total shares into that number, the company would be worth on a silver per share basis of over $200 a share. If you can buy a stock for $37 that is actually worth at least $200, then you have an undervalued stock. Always find out the total value of the metal/mineral reserves since that will give you a strong idea of the company's potential worth, especially in a bull market.

Management that you can dig

The management team should be loaded with mining professionals with at least a decade or more of industry experience. This is information that is usually readily available at the company Web site or industry Who's Who directories. They should have had successful top-management positions with prominent, recognizable mining companies in their career. Professionals that lead mid-size companies and/or have developed a track record of successful mining projects/properties are also desirable.

Using bucks to find ounces

Companies are already producers, majors or top tier companies should already be profitable, and there's no reason you can't find companies with the financial parameters covered in this chapter.

Smaller companies such as explorers and early stage drillers do not necessarily have to have a profit to be considered but financing is important. For them, analyze their financial statements to determine what cash they have on hand and what type of cash flow they generate. They don't necessarily need to be producing a profit now (remember . . . this is speculating) but they need to be involved in projects that will produce a profitable payoff later on.

Politics — Not in my backyard

Where the company's mining properties are located (the political jurisdiction) is very important. On the global scene, lots of mining is being done in countries that have an unfriendly or hostile political climate. Many mining firms were victimized by socialist or authoritarian governments that changed the rules on them. Some of these dictators imposed draconian steps such as arbitrarily raising taxes to exorbitant levels or just outright confiscating the mining company's property (a practice referred to as nationalizing). This can be dangerous for the mining company and, of course, the shareholders as well. There are companies that have seen their stock fall by 50% or more in a single day!

Investors need to keep informed about the politics involved. They need to ask what countries are risky and does the company have much exposure there. Political risk is very real but investors and speculators can add a measure of safety by making sure that they invest in companies that predominantly (or exclusively) are in mining-friendly jurisdictions such as the U.S., Canada, Australia, Mexico, and a few other countries. More about political risks in Chapter 19.

Mining Stocks — Digging Deep

The world of mining stocks is very varied. There are choices that are conservative as well as aggressive. There are some that are appropriate for the stodgy investor as well for the firebrand speculator. You just need to know the difference.

Mix and match

Diversifying is more that just buying a bunch of different stocks. In the world of precious metals, it is more than just a bunch of different mining stocks. It also means understanding yourself and your goals, and being aware of the different kinds of mining stocks. You'll read about the three basic categories of stocks. From an investor's point of view, you can reasonably assume that the first category (the majors) is conservative, the second is growth (the mid-tier, development companies), and the third is speculative (exploratory companies). Depending on the amount you are willing to invest/speculate with, consider 5 to 15 companies across the categories. If you are conservative then by all means have more of your choices tipped in favor of the majors and the development companies.

If you put 100% of your portfolio money into mining then you are getting too speculative so it's a safer idea to have a large portion of your money elsewhere, such as energy, consumer staples, and so on.

The majors

Among mining stocks, this is the safest bet. Besides the fact that they are large companies, they are producing companies. This means that the metals/minerals are already being extracted from producing properties and being sold.

Some examples of large gold mining companies that are producers are Newmont Mining (NEM), Anglogold Ashanti (AU), Freeport McMoran (FCX), and Barrick Gold (ABX). Although they are referred to as gold mining companies, they do mine in the course of their business other metals/minerals that are found in the same mines.

There are fewer primary silver producers but among the major ones are Pan American Silver (PAAS), Apex Mining (SIL), and Coeur D'Alene Mines (CDE). I could probably include a mining stock that is referred to as a land bank, which is Silver Standard (SSRI). A company referred to as a land bank means that it has properties with provable reserves but hasn't mined them. The potential is to ultimately either sell the property to a producer or development company or lease it to others for mining purposes.

The large producers are listed on the major stock exchanges and you can find them and other mining companies at the stock exchange's Web site and do a key word search. In addition, Web sites such as Yahoo! have great stock screening tools that let you find stocks by industry, size (such as market capitalization), and key word.

This top tier is the most appropriate category for conservative investors considering precious metals for their stock portfolios.

Development companies

This category is also referred to as the mid-size or mid-tier companies. They may very well have producing properties but they are primarily developers. They may even do some exploratory activity. Some mature into producing companies. Some examples include Gold Corporation (GG), Agnico-Eagle Mines (AEM), and SeaBridge (SA). They may operate properties that have proven reserves that they are ready to take into the production phase.

This middle category is appropriate for growth-oriented investors. It can be aggressive but not as speculative as the lower tier of small miners.

Eureka! Exploratory companies

This is the category for those looking for the home run. The problem is that you get lots of strikeouts too. In this category you have the smaller companies that are drilling and exploring on properties that may or may not prove to be valuable. This becomes the crapshoot for those who want to speculate. It's important to look for companies (and management) that have a proven record for successful projects.

Unless you have the contacts and the time to do exhaustive research (like visiting the companies and their properties) it is best to get the guidance of experts who extensively cover this corner of the mining world. Some names that come to mind are Jay Taylor, Doug Casey, David Morgan, and Lawrence Roulston. These individuals have newsletters and Web sites so they are easy to find on the Internet, but more importantly they have a long track record and the experience necessary to help speculators in the world of precious metals.

Ancillary companies

Mining is more than just gold, silver, and platinum. It is also more than just U.S. companies. Check out some other places for your money to diversify your portfolio. Consider

- **Conglomerates:** These are mining firms that have diversified properties (with different metals) in diversified locations (different countries). Rio Tinto, BHP Billiton, and Xstrata are good examples of diversified firms that don't put all their metal eggs in one mining basket.

- **Base metals companies:** There's money to be made beyond precious metals. As the global economy keeps growing, the demand for base metals will grow as well. There are profitable companies mining copper, nickel, zinc, aluminum, and other base metals.

- **International opportunities:** Due to globalization, many mining companies based in resource-rich countries such as Canada and Australia are available to U.S. investors through your brokerage account.

Getting Down to the Nitty-Gritty

Every industry has specialized concerns and the mining industry is no different. The followings sections provide you with some points to ponder.

The upside to mining stocks

The bulk of the upside to mining stocks is certainly focused on capital gains. A solid, well-managed company that owns a portfolio of properties with precious metals in a natural resource bull market means achieving fabulous gains that can be legendary. History teaches the best lessons here. Gold and silver mining companies from the late 1970s still count among the greatest percentage gainers in stock market history.

Why? For investors, the secret is that mining stocks offer leverage. If you bought gold at $400 and it gets to $600, you would make a simple investment return of 50%. But what if you took that $400 and bought 100 shares of a $4 gold exploration company? Say it discovers a rich vein of gold at its drilling site that has 2 million ounces of gold. The odds are good that the company's stock price would rise far beyond 50%. That $4 stock could easily double or triple or much more. What if gold goes to $700? $1,000? What would that stock be worth then? You get the picture.

The risks of mining stocks

Any company can run into problems. If you have followed the financials at all in recent years you get a good idea. Most companies go bankrupt for a variety of reasons with the #1 reason always being: not enough money. Among the surviving companies, you can run across lots of rough patches such as economic slowdowns, regulatory problems, too much debt, not enough sales, lawsuits, yadda yadda. This you knew. But every industry or sector has unique negatives and positives as part of the territory. You know about the positive stuff (that's what drew you to begin with). But for the sake of completeness, here's a rundown of pitfalls in the mining industry:

- Mineworkers strike
- Inflationary production costs
- Rising costs of raw materials and energy
- Regulatory changes
- Environmental problems (regulations, protestors, and so on)
- Mining accidents
- Political changes/upheaval (especially in unfriendly countries)

If you have a mining stock and you are concerned about its prospects or unsure about economic conditions facing your stock, consider making some defensive moves. Here are some considerations . . .

- **Protective put:** This strategy means buying a put on your stock. It becomes a cheap form of insurance. Get more information on puts in Chapter 15.

- **Stop-loss orders:** If your stock is at $20 per share, consider putting a stop-loss order at a price of, say, $15 and make it a GTC (good-'til-cancelled) order. If the stock rises, you can cancel your stop-loss order and replace it with a new stop-loss order at a higher price. Your brokerage firm can give you more details about this.

Mining properties

At the crux of the mining company's success is the property to be mined and how productive that property is. A company will either refer to a property's resources or its reserves. There is a crucial difference. When a property has resources that means it may have metals/minerals in the ground but it is not confirmed about the quality and quantity. Reserves is a higher standard. Extensive and reliable testing took place and the company is assured of the quantity and quality of the metals/minerals in the property. The bottom line here is you want to be sure the company has reserves.

When you are researching the company, here are the types of properties that are most desirable:

- **Producing properties:** Properties where high-grade deposits are being mined and the metal is extracted and sold.

- **Development properties:** This property has proven and extractable reserves verified by a reliable feasibility study.

- **Pre-development properties:** Preliminary drilling indicates it has reserves and is ready for a feasibility study.

Any other type of property is purely speculative. If it's referred to as dormant, dead, or speculative, then you are better off leaving it to a man named Jed Clampett and just moving. However, remember that the locations of these properties are very important so read the section entitled "Politics — Not in my backyard," earlier in this chapter.

Hedging practices

Although the general practice of hedging is an honorable one and very necessary for those of us who seek to minimize risk in our portfolios, it is quite a different animal in the mining industry. It can be in fact, a dangerous practice.

Hedging is a practice in the mining industry whereby a mining firm sells, say, next year's mine production this year to lock in a price. Presume that Gopher Gold Corp. sells 1 million ounces of gold today that will be mined next year (also referred to as forward selling). Say that today we sell that gold for $600 per oz., which means we generate current income of $600 million. Some miners do this to lock in income today without worrying about next year's production. Of course, Gopher Gold will have to go to the expense and effort next year to cough up those million ounces of gold that will be due to those folks who paid way in advance. If gold stays at $600 or goes down, then these forward sales look like a smart move. But . . . what if gold goes up?

If gold goes up to $690 an ounce, you have now forgone potential profits of $90 million. Hedging or forward sales are dangerous in a rising market. There have been mining companies that went bankrupt due to excessive hedging. Understand that for a mining company, significant hedging is tantamount to betting against your own asset. This is why many mining analysts prefer unhedged mining companies because in the event the underlying asset (gold, silver, and so on) is rising, the company (and the shareholders) get to fully participate on the upside. Can you visualize that lone protester at the annual shareholders meeting shouting hedging is for gardeners!?

Resources

Because mining stocks can be a specialized area, you need ongoing information and analysis from reliable sources. Here are some of my favorites:

- ✔ Resource Investor (www.resourceinvestor.com)
- ✔ Jay Taylor (www.miningstocks.com)
- ✔ Gold Eagle (www.gold-eagle.com)
- ✔ Kitco (www.kitco.com) and its companion Web sites such as www.kitcocasey.com, www.kitcosilver.com and www.kitco basemetals.com
- ✔ Howe Street (www.howestreet.com). They have extensive reports along with audio and video interviews with industry experts.
- ✔ Korelin Economics Report (www.kereport.com). The highlight is the Internet radio program with audio interviews with experts.
- ✔ Mine Web (www.mineweb.com)
- ✔ Le Metro Pole Café (www.lemetropolecafe.com). Run by analyst Bill Murphy.
- ✔ The Bullion Desk (www.thebulliondesk.com)

Boosting Your Returns

Merely buying and holding mining stocks for future capital gain can be profitable but you can find ways to boost the profitability in your portfolio and squeeze more return on your investment.

Generating income

Until your mining stocks go through the roof 'cause gold zoomed to $3,000 an ounce and the other metals hit nose-bleed territory, you can do something to make more money on your portfolio. The following sections show you a few ways to do it.

Dividends

Normally, mining stocks are not considered income stocks. An income stock refers to a stock that pays dividends greater than the average stock. Lately this means a dividend yielding 4% or higher. Most mining stocks are not income stocks but there are some that issue regular dividends. Some of the large firms do provide a regular dividend (such as Newmont Mining and Gold Corp.). Also, just because the stock has a low dividend, that doesn't mean that it stays there. As company revenues grow, the dividend tends to grow as well. Studies have shown that dividends rise faster than the rate of inflation. That's another plus for long-term investors.

In addition, during the heydays of gold mining stocks (the late 1970s), many in fact issued dividends that very briefly qualified them as high-income stocks. Should gold and other precious metals hit new highs in the coming years then it is very possible that some high dividends can be paid again. Stay tuned!

Writing covered calls

A very easy way to generate income from a portfolio of mining stocks is to write covered calls on your stocks. It is in fact not difficult to safely generate income through this specific strategy of 10% to 20%. Imagine getting income that could be four or five (or more) times greater than a bond or certificate of deposit starting with a mining stock that issues no dividend! It's like being in the twilight zone. Aside from the usual risk of stock investing, covered call-writing is virtually risk-free. To get more information regarding this income-generating approach, look up covered call-writing in Chapter 15.

Leveraging with warrants

Warrants can be an unusual security and most investors are not familiar with them. Technically, they trade like stock (bought and sold like stocks covered in this chapter) but essentially they work like options (covered in Chapter 15). Sooo . . . they warrant coverage here.

A warrant is a financial vehicle that gives the investor the right, but not the obligation, to buy the underlying security (usually common stock) at a specific price during the life of the warrant. A warrant is a wasting asset which means that it will expire on a specific date in the future. However, usually warrants have a long shelf life; freshly issued warrants could expire in as little as two years or as long as five years (or longer). You can say that a warrant can be a hybrid of both a stock and an option. Warrants can be private financial instruments but of course there are many that are publicly traded.

The holder of a warrant (much like the holder or buyer of an option) can exercise the warrant to buy the underlying security but what you can buy differs somewhat with the option. When you buy the typical option, a single option contract represents 100 shares of that particular stock (or ETF). A 2 option contracts means 200 shares and so on. A warrant, on the other hand, is not rigidly standardized the way publicly traded options are.

A typical warrant would have a ratio of one warrant to one share of stock. In other words, if you have 100 warrants of Groundhog Mining Co., this gives you the ability to purchase 100 shares of Groundhog Mining's stock. However, there are plenty of options that are not so cut and dry. There are warrants where the ratio is 1 warrant to 2 shares of stock or to 1 share of common stock and 1 share of preferred stock and so on. Check with the issuer of the warrant (in this case, Groundhog Mining) or the company's shareholder service department.

Before you invest in warrants, get the full details. The most common place to find publicly traded warrants is Nasdaq's Over-The-Counter (OTC) small company market, and information can be found at the OTC Bulletin Board (OTCBB; its Web site is www.otcbb.com). Don't invest in a warrant just because it's available. Instead, do your research on mining stocks and then contact the company's shareholder service department and ask if that particular company has or will issue warrants. Frequently, the details are available at the company's Web site.

Keep in mind that you can also find warrants at the New York Stock Exchange (www.nyse.com) and the American Stock Exchange (www.amex.com), but your most successful searches for metals and mining-related warrants will most likely be at the OTCBB and some international exchanges. Some obvious ones will be in Canada, such as the Vancouver Stock Exchange and the Toronto Stock Exchange. Fortunately, many foreign stocks and warrants do trade on the OTCBB. At the OTCBB, you will notice that foreign securities will be assigned a fifth letter in their symbol, the letter "f," to signify a foreign security.

How profitable can a warrant be? Let me give you a real-life example of how the warrant can work. Silver Wheaton (SLW) is a great example of a major silver mining firm (please don't take this as advice; it is purely an educational example) that also has warrants. One of its warrants (symbol: SLWBF found at the OTCBB site) gives you the ability to buy SLW at $10 per share until the expiration date in late 2010. In late-April 2007, you could have bought SLW for $10 per share and you could have gotten the warrant SLWBF for about $4.50 per warrant. By mid-July, SLW hit $14 while the warrant hit $7.30. How did you fare? During that very brief period (about 3 months), the stock rose 40% while in that same time frame, the warrant went up 62%.

On an annualized return basis, the stock SLW in that time frame shot up 160% while the warrant went up an eye-popping 249%. If a stock doubles or triples (higher?) then the corresponding warrants would perform even more spectacularly. For aggressive investors and speculators that definitely warrants attention.

Another thing that warrants attention is the risk. If the company's stock falls, then the warrant's price would fall and it would likely fall more so. Remember that a warrant (like an option) is a derivative. It is a fancy term which is essentially very simple. A derivative means that this particular financial vehicle or instrument derives its value from something else, the underlying asset.

A warrant or option on Groundhog Mining Corporation would derive its value from the price and performance of Groundhog's common stock. Actually, when you think about, its common stock is a derivative too; it derives its value from the operations, assets, and profitability of the company itself. Stretching it a bit further, a gold mining company would derive its value from the value of the gold it owned or controlled. All right . . . I'll stop at this point before you derive a headache (aspirin is a derivative of tree bark). The bottom line is that you wouldn't analyze the warrant to understand its potential; you would analyze the underlying asset (the company and the metals involved).

Lastly, the most important risk with warrants is the major risk that is also present with options; they can expire worthless. You must do something with them before they expire (such as sell them or acquire the underlying security). Fortunately, unlike most options, warrants do have a long shelf life and that longer time frame does make them less speculative in comparison.

There are very few sources of information on warrants but one that I like is Dudley Baker's Web site at www.preciousmetalswarrants.com. Discuss warrants with your financial advisor. I wouldn't be surprised if you end up educating him or her about the topic!

Indexes

Indexes give us a great snapshot of the broad market or of a specific market, industry, or sector.

If you want to invest or speculate on the indexes, you can do it through exchange-traded funds (ETFs) and/or through options. For investors, there are ETFs that closely mirror or track indexes. You can find more information on ETFs in Chapter 13. For speculators and experienced investors, consider options on the indexes (see Chapter 15 regarding the world of options). Further, there are options on ETFs as well. Man . . . are there choices or what? This is cool stuff . . .

I cover the most widely followed indexes in the following list:

- **PHLX Gold/Silver Sector Index (XAU):** This is a capitalization-weighted index created and maintained by the Philadelphia Stock Exchange (www.phlx.com). Their index, the Philadelphia Gold/Silver Sector Index (XAU), is the broadest index of the gold and silver mining sector.

- **Amex Gold Bugs Index (HUI):** I like this name; gold bugs is a reference to those folks who are very enthusiastic about gold and its prospects. This index is called that because it includes only mining stocks that do not hedge (see segment on hedging earlier in this chapter). To be unhedged in gold production is to be very bullish in gold. The American Stock Exchange (Amex)'s Gold BUGS Index (HUI) is a modified equal dollar weighted index of companies involved in gold mining. To find out more about the HUI, go to Amex's Web site at www.amex.com.

- **Amex Gold Miners Index (GDM):** This index is not as widely followed as the XAU or HUI but it is an important index for precious metals investors. The Amex Gold Miners Index (GDM) is a modified market capitalization weighted index comprised of publicly traded companies involved primarily in the mining for gold and silver. The Index divisor was initially determined to yield a benchmark value of 500.00 starting in December 2002.

Chapter 13

Investing in Mutual Funds and ETFs

In This Chapter

▶ Gaining insight on mutual funds

▶ Understanding exchange-traded funds

▶ Using these vehicles to invest in precious metals

*T*here was a great author who started off a book with "these are the best of times, these are the worst of times" or something like that. As I write this chapter, the conditions on the world scene could qualify as the worst of times depending on how you look at it. In 2007, we worry about too much debt in our country, high taxes, terrorism, Paris Hilton, environmental concerns, inflation, retirement concerns, and so on. On the other hand, when it comes to the fantastic panorama of investing choices, techniques, and strategies available today, it can easily qualify as the best of times. This chapter offers a glimpse of the best of times.

There are many ways of investing, speculating, or otherwise participating in the world of precious metals. Mutual funds and exchange-traded funds (ETFs) are yet another way for investors and speculators to get involved. They are actually ideally suited for conservative investors but can also have features traders and speculators can find appealing.

Mutual Funds

For those of you who see the choices in precious metals or precious metals-related investments a little bit too dizzying, I don't blame you. The feeling's mutual. Sooooo . . . why not a mutual fund? A mutual fund is a great way to invest for many small investors and investors in general who are unsure about any specific investment.

A mutual fund is basically a pool of money that is managed by an investment firm. This pool of money (the fund) is an accumulation of money from people like you and me and many other investors. The fund is managed by an investment firm with a particular investment objective in mind. That objective could be growth, income, preservation of capital, or some other objective. The investment firm will put this pool of money to use by buying a portfolio of securities such as stocks and bonds that the firm believes will meet the stated objective. The bottom line for the investor here is that you choose the fund and you leave the task of choosing the investments in the fund to the investment firm.

Advantages of mutual funds

Mutual funds have endured in popularity since the 1960s and for some very good reasons. The number of mutual fund offerings when I first started in business in 1981 was under 800. By 2006, that number exceeded 10,000 choices, not including hedge funds (at first I thought a hedge fund was something for landscapers). The point here is that there is a mutual fund for virtually any financial purpose and investors can choose one that can truly pinpoint their need(s).

Fill a specific need

With mutual funds, you can choose a fund that fits your needs, not vice versa. If you want growth, you can choose a growth fund. From there you have either aggressive growth or conservative growth. The lesson here is that you don't have to fit your needs into a fund and hope that it works out. You can use a fund that comfortably fits into your personal situation.

Diversification

This is the perennial reason that every source in the universe points out as the advantage of mutual funds. Number 1 is that you can choose mutual funds that invest in only stocks or only bonds, or some mix of the two. Then there are mutual funds that will invest in securities such as stocks in a given sector or industry. And yes, Virginia, there are mutual funds that specialize in the world of precious metals and related investments.

The financial commentators on TV are always crowing about diversification for investors to maintain a balance of both growth and safety. This is where mutual funds come into the picture. Because of this pool of money the mutual fund receives from all these investors they have the financial clout to provide great diversification. This is especially true when you have a sector that could be more volatile and is uncertain and much harder for the average investor to choose winners from. The convenience of the mutual funds is you pick the funds and let portfolio managers of that investment firm pick the winners for you. The great thing is they can choose a losing stock and little harm is done because their financial clout gives them the ability to buy 40 or 50 or more different stocks in the same portfolio.

Convenience

Choosing an investment for your portfolio is not easy. It can take considerable research and due diligence that is not only good but also appropriate. For investors who can't (or won't) do the necessary homework, mutual funds can be the answer. Of course, you still have to choose a fund but it's not that difficult once you understand your investment objective and personal profile.

For precious metals investors who are not sure about what to invest in and don't have the time or inclination to do the research and due diligence that is required of finding a good start to invest in, why not just choose a mutual fund? In a mutual fund, the professionals oversee the entire portfolio on a full-time basis and know what they make by sell and hold decisions on the securities every day.

For precious metals investors . . .

Therefore if you're the type of investor who just scratches his head when he thinks about buying gold, the physical or the paper, gold-mining stocks, or precious metals futures, then you may want to consider a precious metals mutual fund. Would you have made some great money? As I mention in a prior chapter, precious metals have done very well in the year 2000 onward. Therefore, precious metals mutual funds have done very well. In fact, precious metals mutual funds were among the best performers in this decade! See Table 13-1.

Table 13-1	Precious Metals Performance		
Type of fund	*Name of mutual fund (and symbol)*	*Investment firm*	*Web site*
Precious metals	Rydex Precious Metals Fund (RYPMX)	Rydex	www.Rydex funds.com
Gold	Tocqueville Gold Fund (TGLDX)	Tocqueville Asset Management	www. Tocqueville. com
Gold	Midas Fund (MIDSX)	Midas Fund Management	www.midas funds.com
Gold	U.S. Global Investors Gold Shares (USERX)	U.S. Global Investors, Inc.	www.usfunds. com
Precious metals and minerals	U.S. Global Investors World Precious Minerals (UNWPX)	U.S. Global Investors, Inc.	www.usfunds. com
Gold, silver, and platinum group metals	Precious Metals Ultra Sector Profunds	Profunds	www.pro funds.com

(continued)

Table 13-1 *(continued)*

Type of fund	Name of mutual fund (and symbol)	Investment firm	Web site
Precious metals	Vanguard Precious Metals and Mining Fund (VGPMX)	Vanguard Group	www. vanguard.com

---

Please keep in mind that Table 13-1 presents a high-profile sample listing from over 50 precious metals available to investors today. Don't necessarily assume that they are a recommendation or the best performing; it is merely a sample. Continue your research at the mutual fund Web sites mentioned in this chapter.

The downside of mutual funds

Now keep in mind that investing in mutual funds is not all peaches and cream (unless you choose a food and beverage sector fund). Mutual funds do have their downside as well. For example the professional management of the mutual funds is a double-edged sword. The good part is that you get the investment decisions of professionals. But . . . the bad part is that you get the investment decisions of professionals. In other words, you may choose a mutual fund but you generally cannot make the decisions inside the fund and must rely on others. I read an industry report some years ago that pointed out that the average age of the average mutual fund manager was under 35. Hmmm . . . I have socks older than that (just ask my wife). Usually that means that you have someone who hasn't had enough exposure to the markets and their ups and downs over an extensive period of time.

Relying on other decision makers

During the stock market's 2000–2002 bear market, many mutual funds lost money needlessly because the money managers who ran these funds graduated from college during an extensive bull market (the stock market's powerful rise during 1982–1999) so they were clueless about how to manage money in a bear market. The end result is that millions of investors including mutual fund investors lost trillions of dollars. The word to the wise here is simple; make sure you're choosing mutual funds that have been around a very long time and that have experienced both bull and bear markets (and that goes for their financial advisors too!).

Fees and service charges

Another drawback to mutual funds are the various fees involved that investors are assessed through the mutual fund itself. There are two types of fees to be aware of: management fees and marketing or sales fees. Every mutual fund has management fees; obviously you have to pay the people who manage the funds as well as pay for the expenses to maintain the fund such as office and utilities, research costs, and so on. The marketing or sales fees are different matters since they get paid out to those who are actually marketing the fund. The marketing fees that you pay have no bearing on the performance of the fund but they do have a bearing on the financial profitability of the fund. If, for example, you get into a mutual fund that has a 5% front-end loader and that year the mutual fund was up 10%, the real net gain to you as the investor is 5%.

Please understand that I don't necessarily think of fees as a negative. After all, people should be paid for service, performance, and so on. I list them here just as a heads-up so that you understand the cost of investing. Some charge high fees; some charge reasonable fees. As an investor and consumer, you should be aware of all fees involved since they will ultimately have an impact on your investment return. The following sections run down the fees you should be aware of. . . .

Management fees

As I mention, all mutual funds have management fees. These fees cover the operating expenses of the fund (research, office, utilities, and so on) and payment to the professionals running the fund. The amount of the management fee will be expressed as a percentage of the fund and it will vary according to the mutual fund category.

Growth funds, for example, charge a higher management fee usually than, say, a money market fund. A growth fund could charge a management fee of 1.5% to 3% and possibly higher while a money market fund would have a management of under 1% and usually under .5%.

There are also some investment firms (such as hedge funds) that charge a performance fee, which is technically a management fee since it is directly tied to how well the fund is managed. Performance fees can vary widely but most firms charge in the neighborhood of 10% to 20%.

As a general rule of thumb, the more involved, complicated, and aggressive the portfolio, the greater the management fee. Always check the prospectus. The Securities and Exchange Commission (SEC) issues guidelines to make it clear to investors what the costs of investing are for that fund.

Marketing fee #1: The front-end load

The second category of fees are the marketing and sales expenses; referred to as the load. This is paid to the people or organizations that are marketing the fund for the investment firm. Sometimes the load is paid in-house if the investment firm has its own sales force. The first type of load is the front-end load. The load is typically expressed as a percentage of the investment amount.

The front-end load is a fee that is paid in the beginning. If you invest $1,000 with a fund that has a 5% front-end load, then the initial value of your investment is $950. If the investment firm tells you the fund has gone up 15%, don't assume that your investment is now worth $1,150. Your net gain (after factoring in the 5% load) is really up only 9.25% since you have to remember that your investment started at $950.

All fund fees are paid from the fund; no one will send you an invoice. If anyone calls you, you won't have to say the check is in the mail.

Marketing fee #2: The back-end load

The back-end load is a fee that might actually be good for you since it encourages long-term investing. It is a fee that is charged at the time you get out of the fund. It is also called a redemption fee or charge. Most back-end fees are staggered; the longer you are in the fund, the lower the fee is. Many funds actually wipe out the fee once you pass an extended time.

A typical back-end load might be structured as 5% if you are in the fund for less than a year, 3% if you are in one to three years, 1% if you are in three to five years, and 0% after five years. Again, double-check the prospectus for the specifics.

Marketing fee #3: The 12(b)-1 load

The 12(b)-1 load (named after the section in the securities regulations that allow such a charge) is usually a small percentage that is charged every year for as long as you are in the fund.

Say you are in a fund with a 1% 12(b)-1 load. If you are in that fund for a year, your charge would be 1%. If you are talking a $1,000 investment amount, then the charge would be $10. If you are in the fund for 5 years then over that time span you would have paid 1% per year or a cumulative amount of 5%. It may be a small fee but it can erode the return on your investment if you happen to be in that fund for an extended period of time.

Questionable track records

How often do you see that mutual fund ad that raves "The Fantabulous fund was up 47% last year," or "If you invested in our fund when it started ten years ago, today you could buy Rhode Island with enough change to fill your gas tank!" or "Our ultra-growth fund was up 97% just before breakfast." Well . . . some of it could be true but more likely there was some tweaking of the numbers. It's important to put any numbers in perspective.

For example, take the top-performing fund of 1979: a gold fund. 1979 was a hot year for gold as it rose through the late '70s and hit its high in January 1980. During the last five years of the 1970s, the U.S. Gold Services Fund soared over 1,000%. In 1980 and 1981, gold collapsed and gold-related investments plummeted. Guess what? In 1982, that gold fund was still in the top 20 performing mutual funds even though it was at the bottom of the heap for 1980 and 1981. How could that be? When you add up 3 ultra-fantastic years with 1 bad year and 1 horrible year, you still come up with a high, positive yearly average. Energy sector funds had a similar run since oil shot up during the late 1970s and collapsed during the early 1980s.

The lesson is that you have to analyze track records in context and understand the effects of mega-trends. Both bull and bear markets tend to last years so it is important to stay alert when trends stall, sputter, or change directions.

Keys to success with mutual funds

No, this isn't a book or a chapter on mutual funds per se; but there are things to look for when you are looking at mutual funds in the metals and mining sector (and yes . . . some of these points apply no matter what kind of mutual fund you will be investing in). The following sections give you some points to keep in mind.

Analyze yourself first before you analyze the fund

If you are conservative with your money and can't stomach risk and volatility then spread your money over several funds that invest in larger, established companies. In the same way there are aggressive and conservative growth funds in the generic mutual fund realm, there are aggressive and conservative choices in the world of metals and mining. Mutual funds can buy the stock of, say, gold mining stocks that are large, major firms in the industry. Mutual funds can also buy the stock of smaller, more speculative mining companies. (Mining stocks are covered in Chapter 12).

How you invest is as important as what you invest

A famous investing genius once proclaimed stocks fluctuate. Man . . . I wish I said that! Anyway, investments, no matter how well chosen, don't move in a straight line up. The same way bad investments don't go down in a straight line. Single investments, or portfolios, or entire markets tend to either zig-zag up or zig-zag down. Good investments that occasionally pull back (that's called a correction; don't ask me why) present you with a buying opportunity. Bad investments that occasionally go up (a bear market rally) give you a chance to cash out at a better price before they resume declining. Since the world of precious and base metals is in the midst of a long-term bull market, any pull-back provides the opportunity to add to your position for further gains as the up-trend in the market unfolds. (That is . . . if you agree with my prime contention in this book that we are indeed in the midst of a historic bull market for gold, silver, and other metals).

How would you profit on the upward zig zag of the market with a mutual fund? There are three ways.

- **Reinvest the distributions:** When you initially get into a mutual fund, you are asked (in the fund application) to decide what to do with the distributions. Distributions are what the portfolio generates, such as interest, dividends, and/or capital gains. Any and all income or capital gains generated should be totally reinvested. Numerous studies have shown that total reinvestment yields the greatest long-term gains (versus having the distributions sent to you or partially reinvested).

- **Dollar-cost averaging:** On a regular basis, or on pull-backs (or both) send in money to your fund. When the fund's shares are cheaper during those temporary corrections, you get a buying opportunity.

- **Both!** Why not? If you are indeed bullish then consider both total reinvestment of distributions and adding to your fund as opportunities present themselves.

The Prospectus: Netting it out

Yeah . . . I know. You'd rather wait for the movie. Fortunately it's not that bad. Here is a quick way to get through the mutual fund prospectus with relatively little pain. You gotta read before you invest, right? As my uncle Yorgi would say, an ounce of prevention is worth a cheap suit or something like that, so read this before you dig into the prospectus:

- **What is the objective of the fund?** You will usually find the fund's objective right on page 1. In a brief paragraph, you'll know immediately if the objective is yours as well. If not, fuggetaboutit! If the objective is indeed your objective, read on.

- **What does it cost to invest in that fund?** The prospectus should clearly spell out the fees and charges. The less you pay, the more you'll gain (all things being equal).

- **What is the fund invested in and what do they plan to invest in?** A snapshot of the portfolio usually accompanies the prospectus. Now you can see what is in the portfolio and how diversified it is.

- **How has it done in the past?** What is the fund's record and how well did it do not only last year but also over the past three and five years. If you are a long-term investor and the fund has a long history, check out how well it did over ten years.

- **What services does the fund provide?** Ask how they can handle distributions and can they do things such as automatic transfer of money from your checking account for purposes of dollar-cost averaging.

Now if you want to also ask them, "What's the capital of Bolivia?" well, that's up to you. Now if you don't follow all these great tips and then make a mistake with mutual funds, don't complain. Like my Uncle Yorgi would say, "You spoiled the broth . . . now you lie in it!" Man . . . I wish I said that.

Mutual Fund Resources

I can't tell you everything I want to write but there are some great places to turn to for extensive information on mutual funds in general and in the great area of metals and mining. Here are some great Web sites for mutual fund investors:

- ✔ Investment Company Institute (www.ici.org)
- ✔ Mutual Fund Education Alliance (www.mfea.com)
- ✔ Best No-Load Funds (www.bestnoloadfunds.com)
- ✔ MutualFunds.com (www.mutualfunds.com)
- ✔ Morningstar (www.morningstar.com)
- ✔ Value Line (www.valueline.com)

Exchange-Traded Funds

I love exchange-traded funds (ETFs). Really! I think that they are the best financial innovation in the past five years. The investing public must think so too; nearly $500 billion had been invested in ETFs by the end of 2006.

First of all, what is an ETF? An ETF is like a mutual fund but it trades like a stock. It's like a mutual fund in that it has a basket of securities that it maintains but the similarity ends there. The basket of securities in the ETF is typically a fixed basket of securities while the mutual fund's securities are an actively managed portfolio. In other words, the mutual fund's basket of securities (such as stocks, bonds, and so on) is not static; the securities are regularly bought and sold as well as held.

The ETF also trades like a stock, another difference it has with mutual funds. When investors buy or sell shares of a mutual fund, they deal directly with the fund. The ETF, on the other hand, is bought and sold like stock. Just like a stock, you can buy 1 share or 1,000 easily through any stock brokerage account. In the following sections, I give you an overview of ETFs, specifically, precious metals ETFs.

The pros and cons of ETFs

ETFs have the following advantages:

- ✔ **Diversification:** Depending on the industry, the ETF could have 20 to 50 different issues in its portfolio. Most ETFs tend to focus on a specific industry or sector.

- ✔ **Personal control:** You decide how many shares you want to buy. You decide whether to hold or sell. You can even use brokerage orders such as stop-loss orders or virtually any brokerage order that applies to a stock.

- ✔ **Tax advantages:** Because the ETF's basket of securities is a relatively fixed portfolio, it is tax efficient in that the individual securities are seldom sold.

- ✔ **Liquidity:** Just as with a stock, you can buy or sell it very easily.

- ✔ **Flexibility:** You can dictate how many shares and be able to buy the right amount in your account whether you have a large or small account.

- ✔ **Lower costs:** ETF fees are relatively low and a small percentage of the portfolio (especially when compared to mutual funds with high loads).

- ✔ **Ownership benefits:** You own an ETF as though it's a stock; you can use it as collateral (ETF's are marginable) and you use it to write covered call options (options are covered in Chapter 15).

Those are great benefits. Just thinking about them makes the tears well up in my eyes. Ok . . . let's get back down to earth. Here are some of the drawbacks:

- ✔ **The shortcoming of diversification:** Diversification reduces risk but it can also reduce gain. A singe stock may have great risk but if you are right about it, it has great potential for gain. Yes . . . there are trade-offs.

- ✔ **Industry or sector risk:** If you choose an ETF in an industry that is having a rough time, it will cause the ETF's price to fall.

There is a unique risk with ETFs that specialize in the metal itself. ETFs such as the Gold ETF (symbol: GLD) and the Silver ETF (symbol: SLV) do have a question about whether they actually have the metal itself.

The world of precious metals ETFs

Because ETFs are an excellent way for investors and speculators to participate in the precious metals arena, here are all the ETFs (so far!) that you should be aware of . . .

The gold ETFs

Investing in gold gets easier and easier. These are the primary gold ETFs in order of trading volume (with their trading symbols):

- streetTRACKS Gold Shares (GLD)
- iShares COMEX Gold Trust (IAU)
- PowerShares DB Gold Fund (DGL)

The gold ETFs act like the metal but in paper form. In other words, the gold ETFs allow you to get involved in gold but as a paper investment that you can transact much like a stock. You see, you can't normally (or at least conveniently) buy physical gold through a stock brokerage account (or through an IRA if it is in, say, a self-directed brokerage account or a mutual fund). But with the advent of gold ETFs, you can now participate in the gold market conveniently and as easily as any stock or bond.

The silver ETFs

For those who are excited about the potential profits in silver (like me) then these silver ETFs are worth considering:

- iShares Silver Trust (SLV)
- PowerShares DB Silver Fund (DBS)

Here again is the same point as with the gold ETFs. The silver ETF gives investors and financial institutions the ability to include silver as an investment choice inside the realm of a brokerage account or IRA.

Other precious metals ETFs

The following are metals-oriented ETFs that add a level of diversification. They have more than just one metal:

- PowerShares DB Precious Metals Fund (symbol: DBP). This ETF concentrates on the two most prominent precious metals, gold and silver.
- Central Fund of Canada (symbol: CEF). Technically not an ETF but has all the characteristics of one. CEF has gold and silver physical bullion reserves.

Other ETFs for the metals-minded

For the sake of completeness, I want to include some ETFs that have metals as an important component but that have other, non-metal components as well. I do this because investors should be aware of other securities that are not necessarily a pure-play involving 100% precious metals. The reason is that an ETF can have diversification for those investors who want to hedge their bets, so to speak, about precious metals.

- ✔ Market Vectors Gold Miners ETF (GDX). This ETF mirrors the 30 gold mining indexes found in the AMEX gold mining index.

- ✔ PowerShares DB Base Metals Fund (DBB). This fund is based on an index of the most commonly used and most liquid base metals, such as aluminum, copper, and zinc.

- ✔ PowerShares DB Commodity Index Tracking Fund (DBC). This fund reflects an index of six of the most heavily traded commodities in the global marketplace: gold, aluminum, crude oil, heating oil, corn, and wheat.

- ✔ SPDR S&P Metals & Mining ETF (XME). This ETF mirrors the S&P metals and mining index which is composed of 30 companies involved with precious and base metals and mining.

All the above ETFs have generally performed well during the past few years of the metals' bull market. If you're looking to participate in precious metals with some measure of safety, this is the venue for you. Just ask Uncle Yorgi.

Oops . . . before I forget, I want you to investigate options on ETFs. Some of the greatest profits in recent years for myself and my clients and students have not just been in ETFs; they have also been with options on ETFs. There are a variety of ways to profit with options and there are options strategies suitable for virtually any investor. To find out more about options see Chapter 15.

ETF Resources

As ETFs proliferate, so will the information and opinions about them. Here are among the best resources to help you choose. I'm sure you'll see a lot more coming out of the woodwork in the coming months.

The exchanges

Most of the ETFs you'll come across will most likely emerge on either the American Stock Exchange (www.amex.com) or the New York Stock Exchange (www.nyse.com). As of this writing, the American Stock Exchange has more information and resources for ETF investors since it has been a trailblazer in this hot area. You'll find lots of details on ETFs and you have the ability to download prospectuses (in Adobe PDF file format).

The issuers

Since the ETF universe keeps expanding, you should get to know the major firms that issue ETFs. These firms do the heavy lifting of creating the ETFs, listing them on the stock exchanges, and maintaining them. See Table 13-2 for an extensive listing of ETF issuers.

Table 13-2	ETF Issuers	
The ETF series	**The financial firm that creates and maintains it**	**Web site**
iShares	Barclays Global Investors, N.A.	www.ishares.com
PowerShares	PowerShares Capital Management, LLC	www.powershares.com
Claymore	Claymore Securities, Inc.	www.claymore.com
Rydex	Rydex Investments	www.rydexfunds.com
First Trust	First Trust Portfolios, L.P.	www.ftportfolios.com
SPDRs	Standard & Poor	www.sectorspdr.com
Ultra	SEI Investments	www.seic.com
Market Vectors	Van Eck Securities Corporation	www.vaneck.com
HOLDRS	Merrill Lynch & Co.	www.holdrs.com
Street Tracks	State Street Bank & Trust Co.	www.statestreet.com
Vanguard	Vanguard Group	www.vanguard.com

Peruse the extensive list of ETF issuers and their roster of ETFs. More are added constantly and due to the boom in natural resources in general and metals in particular, you have (and will have) a great selection of precious (and base) metals–related ETFs.

Investors are best served if they stick to ETFs that are very liquid. (Yes, there is a water ETF but that's not what I mean!) In other words, stick to ETFs that have a large market (many buyers and sellers) so that it is easier to buy and sell your ETF. Judging the liquidity can be as easy as looking at the trading volume. (Most active ETFs have a daily volume of over 50,000 shares.) Low liquidity or low activity sometimes results in unfavorable prices due to matters such as a wide bid/ask spread or a difficulty to quickly fill an order.

Newsletters

The ETF part of the financial landscape is here to stay and it will continue to grow. Their limitation is only set by the imagination of the issuers. You can find an ETF for almost any purpose

- ✔ ETF Investment Outlook (www.investmentoutlook.com)
- ✔ ETF & CEF Newsletter (www.etfinvestornewsletter.com)
- ✔ ETF Watch (www.etfwatch.com)

The large publishers such as Forbes, Weiss Research, and Value Line have newsletters and research reports that cover ETFs (and mutual funds) with in-depth data. Well-stocked business libraries carry their publications. And of course, use a major search engine to find them and other sources.

Web sites

Fortunately, the same general financial information Web sites that tell you about the economy in general and the financial markets in particular will very likely have lots of information on ETFs (and mutual funds) and how each one is faring lately. Here's some to start you off:

- ✔ Market Watch (www.marketwatch.com)
- ✔ Bloomberg (www.bloomberg.com)
- ✔ Yahoo! Finance (www.finance.yahoo.com)
- ✔ ETF Global Investor (www.etfglobalinvestor.com)
- ✔ Investopedia (www.investopedia.com)

Chapter 14

Exploring Futures

In This Chapter

▶ Checking out the history and main players in futures

▶ Understanding the fundamentals of futures

▶ Checking out different futures contracts, investment strategies, and resources

Fasten your seat belts. Check your helmets. Grab some bromo-seltzer. You're now ready for the world of futures! Fortunes can be made quickly. They can be lost even faster. Let's dip our collective toes in the water of this exciting corner of the financial world.

Back to the Futures

Commodities and futures are mentioned as though they are synonymous but, at the risk of nit-picking, they are truly different. First of all, commodities make up the raw stuff that is indeed the building blocks of society. Commodities encompass food, building materials, precious metals, energy and more. Commodities range from the typical (wheat, soybeans, sugar, and so on) to the odd sounding (pork bellies?). Commodities end up in their ultimate forms as produce and processed foods found at the local supermarket or in other venues such as the gasoline you get at the local gas station or the components found in the circuits of a computer or cellphone. They are in every corner of modern life.

Now how about futures? Futures are nothing more than commodities that come in a standardized quantity and are delivered in the future at a specified price and a specified date in the future (hence the term *futures*). The word *futures* is technically short-hand for *futures contract* since futures contracts are really legal claims to an underlying asset (the commodities). The reality then is that commodities are real stuff while futures are paper. So futures contracts are derivatives; they derive their value from the underlying asset. This is a crucial distinction (you'll see why later in this chapter).

A brief history on futures

In the mid-1800s, merchants began organizing centralized markets to make the act of buying, selling, and trading commodities much easier. These places gave them the ability to set standards for quality and quantity. The first commodities to be commonly traded were grains and other agricultural products. By the mid-1900s, over 1,500 markets or exchanges had been set up at every major port and rail station. As America became industrialized, these exchanges went beyond agricultural commodities into almost anything that had an active market.

The major commodity hubs sprung up during the 19th and 20th centuries at two large urban cities . . . Chicago and New York. In New York City, circa 1872, a group of dairy merchants came together to create a formal exchange which came to be called the Butter and Cheese Exchange of New York.

As they added commodities, the exchange grew and so did the name. It became the Butter, Cheese and Egg Exchange. I would hate to think what the local supermarket would have been called then but fortunately, they made it a convenient name in 1882 when it became the New York Mercantile Exchange (NYMEX). A second exchange, Commodity Exchange, Inc., was established during the Great Depression in the early 1930s. This came to be called COMEX and it specialized in hard goods and metals such as copper, silver, and tin.

Because the federal government banned private ownership of gold during the Great Depression, it was not available for futures trading on the COMEX until December 1974. New York's two largest commodities exchanges, COMEX and NYMEX, merged in August 1994.

The two exchanges became two divisions of one entity. The NYMEX Division traded energy (such as crude oil, heating oil, gasoline, and natural gas) and the precious metals platinum and palladium. The COMEX Division covered gold, silver, copper, aluminum, and several stock indexes.

In the early days (the 1800s and early 1900s) transactions were primarily in cash on the spot for immediate delivery (the spot market) and in forward contracts (described later in this chapter) before the advent of futures contracts.

I think that a good way to remember the difference between commodities and futures is somewhat like the difference between a car and a car title. The car is the underlying asset and the car title is a formal piece of paper that has a legal claim to a car. In the same way a futures contract precisely describes what it is a claim to (a certain quantity and quality of a commodity), a car title precisely describes what it is a claim to (a 1987 Ford Torino, its serial number, and so on). This is a very insightful paragraph so I might as well end it here.

Although the futures market may look sometimes like a crazy and haphazard market, the futures market is an orderly market that performs important functions in our overall economy. Although futures are a speculative and risky market for our purposes, it is actually a market that removes risk and uncertain for those in industry that need the underlying commodities to keep making the stuff that you and I need on a daily basis.

How the futures market works

The futures market operates in a centralized marketplace called the futures exchange. The exchange makes sure that the market works smoothly. Those floor brokers are there to match buyers and sellers with the best prices. Each of the stations represents a batch of commodities and that chaotic crowd that hovers around the stations is making buy and sell bids based on the open cry system. In that crowd are brokers who are buying futures for their clients or for their own accounts. Those shouting for the buyers are looking for the lowest prices. Those shouting for the sellers are looking for the highest prices. This method has real and beneficial impact not only for those in futures but also the economy at large.

Two very important events occur on the floor of the exchange with all these brokers:

- ✔ **Price discovery:** Because of how competitive it is, the futures market becomes an important way to determine prices based on current and future estimated supply and demand. The futures market receives a non-stop flow of information from around the world, all of which gets absorbed and ends up being reflected in the price through the bidding process. This is how the price is discovered. This helps the rest of the economy figure out prices and values of an entire range of goods from A to Z. Capitalism at its finest!

- ✔ **Risk reduction:** The futures markets provide a place for commerce and industry to reduce risk when making purchases. Risk is reduced because the price in the futures contract is set which helps those who want delivery of the underlying asset know exactly what they are buying (or selling), how much they are paying (or receiving), and when delivery can (or should) be made. Some examples appear later in this chapter (see hedging).

That's right . . . futures can reduce risk. Your response might be "Reduce risk?! Are you crazy?! I lost a fortune going long umbrella futures just before the drought! How can futures possibly reduce risk?" Glad you asked. Technically, risk is not reduced . . . it is transferred. Merchants and manufacturers seek to reduce the risk in buying their raw materials and the folks who take on this risk for them are the speculators. Speculators take on risk because of the incentive to possibly make a lot of money.

The futures market is a crystal clear example of the age-old equation of risk versus return. Desiring a greater return on your money means tolerating greater risk. If you don't want more risk, then you have to tolerate a lower return on your money (yep . . . it can be a vicious cycle).

What can be traded as a futures contract

The average person would be surprised as to the range of things that can be traded as futures (the average person is also annoyed at being called average). Although we concentrate on precious metals, it's important for you to see the range since many traders do two or more commodities (or other tradeable entities such as currencies and financial instruments). A trader may, for example, take positions in gold and the U.S. dollar to take advantage of opposing price moves. It's important for you to know what's available for trading opportunities or to just plain diversify your futures account:

- ✔ **Precious metals:** The basic ones are gold, silver, platinum, and palladium.

- ✔ **Base metals:** Copper, aluminum, zinc, lead, tin, and nickel (the non-ferrous metals). Find more details on base metals in Chapter 9.

- ✔ **Steel:** A ferrous metal (not to be confused with a Ferris wheel).

- ✔ **Currencies** (also known as Forex which is short for foreign exchange): The most actively traded are the U.S. dollar, euro, Japanese yen, Canadian dollar, British pound, Australian dollar, and others.

- ✔ **Indices:** The S&P 500, E-Mini S&P, Nasdaq, Dow Jones, FTSE, DAX, and others.

- ✔ **Grains:** The most active are corn, wheat, soybeans, soy meal, bean oil, oats, rice, rapeseed, barley, and others.

- ✔ **Financial instruments:** Treasury notes, treasury bonds, muni-bond, eurodollar, BUND, Schatz, Euroyen, and others.

- ✔ **Food and fiber:** Coffee, cocoa, sugar, cotton, orange juice, lumber, and others.

- ✔ **Energy:** Crude oil, Brent Crude, West Texas Intermediate Crude, natural gas, unleaded gasoline, propane, and others.

- ✔ **Meats:** Live cattle, pork, lean hogs, and others.

- ✔ **Plastics:** That includes . . . uh . . . plastics.

- ✔ **Stocks:** There are futures on single stocks.

- ✔ **The unusual:** Hurricane (you read that right), home prices, and who knows what else they'll think up.

The futures market is such a growing area, I wish there were futures on futures (maybe in the future).

The Players in the World of Futures

The main players in the world of futures are the exchanges, speculators, hedgers, and the regulators. Check out the sections that follow for more information on each.

To make all this great futures activity happen, you will need a broker and an account approved, funded, and ready to go. I could write a whole chapter on brokers so I did . . . see Chapter 17.

The exchanges

All the futures trading action has to happen somewhere. That place would a centralized market that makes the buying and selling efficient and liquid; that place would be a futures exchange. Here are the major futures exchanges in the United States:

- ✔ New York Mercantile Exchange (NYMEX): This is the most active exchange for precious metals (www.nymex.com).
- ✔ Chicago Mercantile Exchange (CME; www.cme.com).
- ✔ Chicago Board of Trade (CBOT; www.cbot.com).

Don't forget that these exchanges also offer options on futures contracts. Speak to your broker about options since they are great tools for speculators.

For speculators in base metals (both futures and options), you may need to look outside the United States for opportunities not readily available at exchanges mentioned. A sign of how markets have become global is that now it can be easy to trade futures in your account that are in an international futures exchange. The largest and most active ones are

- ✔ **London Mercantile Exchange (LME):** They offer futures and options on zinc, lead, nickel, tin, aluminum, and copper. Their Web site is www.lme.co.uk.
- ✔ **Eurex:** This is a European futures exchange very active in financial instruments such as derivatives (www.eurexchange.com).
- ✔ **Tokyo Commodity Exchange (TOCOM):** They have precious metals futures contracts (as well as some unusual ones such as rubber and kerosene futures). Find out more at www.tocom.or.jp.

As the world becomes a more global market, more and more exchanges will either spring up or existing exchanges will make available their own offerings in precious metals. Exchanges ranging from Dubai (pronounced "do buy"; hopefully they "do sell") to Shanghai have already approved trading in precious metals such as gold and silver.

Keep in mind that all this international futures trading will benefit you even if you don't buy or sell a single ounce of any precious metal. As more investors across the globe get involved, that will increase demand. Couple this with limited supply and you have yet another bullish factor for precious metals.

Speculators

This is the category that you and I belong to. We don't want to take delivery of 25,000 pounds of copper – we just want to profit from its market action. As market participants, speculators are out to profit from the futures market risk and volatility. Speculators are not involved in some business that needs the goods tied to the futures contract; they are looking for cash transactions. Depending on the market, only 1% to 5% of futures contracts result in a physical delivery. Speculators are not seeking safety but the financial benefits that could accompany the risky side of the market.

Hedgers

Hedgers are not landscapers. That is just a term for those who practice hedging, which is a practice of making an order to reduce the risk of an unfavorable price move in an asset or commodity by taking an offsetting position in a related security (such as a futures contract). The odds are that the hedger is an organization in commerce or industry (such as a manufacturer or an agricultural business) that regularly buys raw materials for its production facilities. Here's an example . . .

Baubles Galore, Inc. (BG), is a gold jewelry manufacturer that must secure some gold before the holiday season to produce earrings and necklaces for retail consumers who are already being marketed in catalogs and Web sites with prices already publicized. Say that season is six months away. What happens if the price of gold goes up during that time? The prices that the consumers would be paying are already set. If gold goes down in that period, not a problem. BG would end up making more money. But the real problem is if prices go up. That is a real risk for BG since it cannot pass along the higher cost to its customers. What can the company do?

Fortunately, the company can hedge or diminish the risk of the potential gold price increase by entering the futures market and purchasing a gold contract that would settle in a suitable time frame (such as August) and lock in a price. Let's say that time has passed and it is now August and the price of gold is up $50 per ounce. Fortunately, BG is not affected by the increase of the price of gold since the futures contract locked in the lower price.

Now you see the value of hedging and how the firms that do take delivery benefit from a futures contract. Locking in a price protects you from unforeseen price increases. Hedging with futures contracts can protect buyers and sellers. Once you understand who has the incentive to lock in a lower price (or a higher price) you can understand the economic value of futures contracts to hedgers. Cattle ranchers would hedge against lower meat prices while a large restaurant chain would hedge against higher meat prices. In each case, the risk was transferred to the other side of those futures contracts, the speculators.

The regulators

America's futures market is regulated by the Commodity Futures Trading Commission (CFTC), a federal government agency (www.cftc.gov). The market is also subject to regulation by the National Futures Association (NFA), a private, self-regulatory body authorized by the U.S. Congress and subject to CFTC supervision. The NFA (www.nfa.futures.org) primarily overseas the brokers and enforces a code of business ethics.

A futures broker (the firm and those individuals employed as brokers) must be registered with the CFTC in order to issue or buy or sell futures contracts. Futures brokers must also be registered with the NFA and the CFTC in order to conduct business. The CFTC has the power to seek criminal prosecution (through the Department of Justice) when it thinks that the law may have been violated. The NFA can permanently bar a company or an individual from dealing on the futures exchange because of violations of business ethics and the NFA code of conduct.

Investors and speculators should become more informed about how futures should work and aware of anything improper. To find out more, head to the Web sites for the CFTC and the NFA. You will find great consumer information on choosing a broker and educational materials for how the market works.

If you feel that you have been victimized by fraud or improper behavior, contact the CFTC and the NFA for assistance.

The Fundamentals of Futures Contracts

At the core of futures trading is the futures contract. The futures contract derives its value from the underlying asset (corn, gold, and so on). In this respect a futures contract is really a derivative. A derivative sounds complicated but the essence of a derivative is that it (a paper contract, for example) derives its value from something tangible.

In a futures contract, the agreement is between two parties. Party #1 agrees to deliver a commodity (referred to as the short position). Party 2 agrees to receive a commodity (referred to as the long position).

After all, what is being traded? The futures contract is a paper vehicle with the underlying promise to deliver the goods, the commodities involved. A buyer of a futures contract may or may not want delivery. It depends on why they are buying into the futures contract. Buying the futures contract is also referred to as going long. There are two basic reasons that a buyer will buy a futures contract:

> The buyer is an organization that may want to see a delivery of the physical commodity made. Maybe the organization is a jewelry manufacturing firm that needs raw materials (such as gold or silver) to make jewelry.

> The buyer is a speculator seeking profit. He or she doesn't want the underlying commodities (who wants a ton of pork bellies or copper dumped on their lawns anyway?); they are just looking for making some money betting on the price move of the underlying commodity.

Features of the futures contract

Like any contract, a futures contract is an agreement. It is a saleable contract and it is made on the trading floor of a formal futures exchange. The contract is for the buying or selling of a particular commodity (in our case, metals) at a predetermined price at a specified delivery date in the future.

A specified size

Futures contracts get standardized for the same reason when you go to the store to get a loaf of Wheat-O brand bread: it is the exact same as any other loaf of the same type.

The futures contract is tied to an asset with a set size. Each security futures contract has a set size. The size of the security futures contract is determined by the regulated exchange on which the contract trades. For example, a security futures contract for a single stock may be based on a hundred shares of that stock. If prices are reported per-share the value of the contract would be the price times 100. For a narrow-based security indices the value of the contract is the price of the component securities times the multiplier set by the exchange as part of the contract firms.

Specified quality

Not just any old batch of stuff can be an underlying asset to a futures contract. The exchange is usually pretty vigorous about quality. Take the Chicago Board of Trade (CBOT) and its silver futures contract. The description of the underlying metal states clearly that it must be "5,000 troy ounces (±6%) of refined silver, assaying not less than .999 fineness, in cast bars weighing 1,000 or 1,100 troy ounces each." You can't just slap on some silver paint to a copper brick and expect it to be acceptable. Buyers and sellers expect a standardized quality.

Specified prices

The prices of security futures contracts are usually quoted the same way prices are quoted in the underlying asset. For example, a contract for an individual security would be quoted in dollars and cents per share. Contracts for indices would be quoted on an index number, usually stated to two decimal points. Each security futures contract has a minimum price fluctuation (called a tick), which may differ from product to product or exchange to exchange. For example, if a particular security futures contract has a tick size of one dollar, you can buy the contract at $100 or at $101 but not at $100.50.

A specified delivery date

Futures contracts on precious metals will continue to be traded right through the deliver money until trading is terminated on a day specified by the exchange near the end of the month. Up until expiration you may liquidate an open position by offsetting your contract with an equal and opposite contract that expires in the same month. If you do not liquidate an open position before it expires, you will be required to make or take delivery of the underlying security or to settle the contract in cash after expiration.

Although security futures contracts on a particular security or a narrow-based security index may be listed in and traded on more than one regulated exchange, a contract specifications may not be the same. Also, prices for contracts on the same security or index may vary on different regulated exchanges because of different contract specifications.

Because each futures exchange has slightly differing terms for its contracts, review the contract specifications at the exchange's Web site. Discuss it with your broker so you're not surprised.

A different animal: The forward contract

Sometimes you may hear the phrase "forward contract" when commodities are discussed. This is not the same as a futures contract. A forward contract is usually a private contract (not a publicly tradeable contract like a futures contract) that is created in a cash market between a buyer and a seller of the asset in question. Although the delivery of the underlying asset is made in the future, the price is settled on the day of the trade. Forward contracts are not standardized nor are they transacted at the exchange.

Metals Futures Contracts

You can find futures contracts for both precious as well as base metals, and you can also get in the futures games with a smaller amount of dough by using mini-futures contracts. I discuss these futures contracts as well as special considerations, such as margin and leverage, that you should keep in mind if you decide to get involved in futures.

Precious metals contracts

There are futures contracts on gold, silver, platinum, and palladium. There are more obscure precious metals (like rhodium) but their markets are very thinly traded and not really practical for the average investor. Table 14-1 gives you a rundown on the main precious metals futures contracts (at NYMEX).

Table 14-1		Precious Metals Futures Contracts			
Metal	*Contract size*	*Delivery months*	*Symbol*	*Tick*	*Point value*
Gold	100 oz.	G, J, M, Q, V, Z	GC, RGC	10 points = $10	1 point = $100
Silver	5,000 oz.	F, H, K, N, U, Z	SI, RSI	50 points = $25	1 cent = $50
Platinum	50 oz.	F, J, N, V	PL	10 points = $5	1 point = 50 cents
Palladium	100 oz.	H, M, U, Z	PA	5 points = $5	1 point = $1

Metal	Contract size	Delivery months	Symbol	Tick	Point value
Uranium*	250 lbs. of U308	*	UX		

Uranium futures started trading at NYMEX in May 2007. These are cash-settlement only contracts meaning there is no delivery of uranium (whew! that's a load off my mind).

For delivery months, the column reads with letters that the exchange uses to symbolize the delivery month. Table 14-2 gives all the months and their accompanying letter (symbol):

Table 14-2	Delivery Month Symbols
Delivery month	*Symbol letter*
January	F
February	G
March	H
April	J
May	K
June	M
July	N
August	Q
September	U
October	V
November	X
December	Z

As an example of how a particular contract is symbolized, a platinum contract for the delivery month October 2008 would be "PL V8." NYMEX's Web site gives you full details on the futures contracts and their delivery dates and margin requirements.

Trades on a futures contract continue through its delivery month until a final date specified by the exchange. Gold futures contracts for December 2008 are terminated for trading as of December 29, 2008.

Base metals contracts

Table 14-3 gives you the most active futures contracts that are traded for base metals.

Table 14-3	Base Metals Futures Contracts		
Metal	*Exchange*	*Contract size*	*Symbol*
Copper	NYMEX	25,000 lbs.	HG, RHG
Aluminum	NYMEX	44,000 lbs.	AL
Aluminum	LME (London)	25 tons (with a tolerance of +/- 2%)	See LME
Zinc	LME (London)	25 tons (with a tolerance of +/- 2%)	See LME
Nickel	LME (London)	25 tons (with a tolerance of +/- 2%)	See LME
Lead	LME (London)	25 tons (with a tolerance of +/- 2%)	See LME

Mini-futures contracts

For those of you who think you need a zillion bucks to get into futures, think again. The exchanges have something for those who would like to speculate with less money, which also means less risk.

Recently, some of these mini contracts had margin requirements of under $500, which makes it more affordable to speculators with limited funds. Of course, check with your broker since initial margin can change.

See Table 14-4 for a list of mini-futures contracts.

Table 14-4	Mini-Futures Contracts		
Metal	*Exchange*	*Contract size*	*Symbol*
Gold	CME	33.2 troy oz.	YG
Gold	NYMEX	50 troy oz.	QO
Silver	CME	1,000 troy oz.	YI

Metal	Exchange	Contract size	Symbol
Silver	NYMEX	2,500 oz.	QI
Copper	NYMEX	12,500 lbs.	QC

Pass the margin

A futures contract is an obligation and not an asset; it has no value as collateral for a loan. Because of the potential for a loss as a result of the daily market to market process, however, a margin deposit is required of each party to a security futures contract. This required margin deposit also is referred to as a performance bond. This performance bond or margin is security for the broker.

The basic margin requirement is 20% of the current value of the security futures contract. Some types of futures strategies may have lower margin requirements. Requests for additional margin are known as margin calls. Both the buyer and seller must individually deposit required margin to their respective accounts.

It is important to understand that individual brokerage firms can, and in many cases do, require margin that is higher than the exchange requirements. Additionally margin requirements may vary from brokerage firm to brokerage firm. Furthermore, a brokerage firm can increase its house margin requirements at any time without providing advance notice, and such increases could result in a margin call.

For example, some firms may require margin to be positive the business day following the date of the deficiency, or some firms may even require deposit on the same day. Some firms may require margin to be on deposit in the account before they will accept an order for a security futures contract. You should get very familiar with the customer agreement with your brokerage firm before entering into any transactions in security futures contracts.

Brokerage firms generally reserve the right to liquidate a customer's security futures contract positions or sell customer assets to meet a margin call at any time without contacting the customer. Brokerage firms may also enter into equivalent but opposite positions on your account in order to manage the risk created by a margin call. Some customers mistakenly believe that a firm is required to contact them for a margin call to the valve and that the firm is not allowed to liquidate securities or other assets in their accounts to meet a margin call unless a less the firm has contacted them first. This is not the case.

While most firms notify their customers of margin calls and allow some time to deposit additional margin you are not required to do so. Even if a firm notifies the customer of a margin call and sets a specified due date for a margin deposit, the firm can still take action as necessary to protect its financial interests, including the immediate liquidation of positions without advance notification to the customer.

For speculators who want to avoid the problems inherent in margin of your account, consider options on futures. Generally, options on futures for retail investors and speculators can be bought with cash and done so without margin. Options on futures have the same basic volatility and can offer great profit potential but can be done without the risks of margin. You can lose money in options but at least the worst case scenario is that losses are limited to the price of the option and no more. See Chapter 15 for more about options. Almost every precious metal and base metal covered in this book has options available. Explore this with your broker.

Leverage: The double-edged sword

The futures market can offer leverage which can magnify your gains (or your losses). Leverage is the ability to retain control over a valuable asset with a relatively small amount of money. Most people are familiar with leverage in real estate; you gain control of a property valued at $250,000 by making a small down payment of, say, 10% ($25,000) and borrow the remaining 90% ($225,000). Then if the property goes up to $300,000 in value (a gain of $50,000), your initial invested amount of $25,000 at that point made you 200%.

In futures, you get similar leverage. On the plus side, you don't have to use as much money as in real estate and you don't have to wait a long time to see results. On the negative side, the leverage can work against you and you can lose money big time. I tell you . . . there's always a catch!

In futures, the down payment is your initial deposit (not really a down payment; just comparing it with the prior example). As in the previous example, the initial deposit or margin is a small sum that in turn controls a valuable asset as represented by the valuable futures contract.

Take a futures contract on gold. The contract's underlying asset is 100 oz. of gold. If gold is $650 an ounce, then the contract would be worth $65,000 (100 ounces times $650 an ounce). Say the initial margin is $5,000 and you are going long on the gold contract. If gold rises to $710, then that contract would be worth $71,000 (100 ounces times $710 an ounce). Since the price rise would be $6,000, your profit percentage would be 120% (the $6,000 gain is 120% of your initial margin deposit of $5,000).

Again, if gold fell in that example, you would lose money as your broker makes that margin call so that you can kick in more funds to maintain margin.

Basic Futures Trading Strategies

The whole point of speculating in futures contracts is an attempt to figure out what the price will be in the near future and profit from that move. Your strategies all flow from what your expectation is. Although there are many sophisticated strategies with futures (especially when you combine them with options on futures), that is something you can look forward to learning after this book. For now, the basics will be covered. Very basic. But some of the best strategies are indeed simple.

Basic strategy #1: Going long

In going long, the speculator is making a bullish bet (thinking that the asset will go up in price) on that futures contract. The speculator gets into the contract by agreeing to buy and receive delivery of the underlying asset at a set price. The bet is that the market price will rise above the set price and the profit is the difference.

Say that Fred the speculator is bullish gold. Fred would therefore go long a gold futures contract. Presume he starts with an initial margin of $2,000 in August. He buys one December contract with gold at $600 per ounce. Since the contract is for 100 ounces of gold for a total contract value of $60,000, and since Fred is going long in August he is expecting (hoping?) that the price of gold will go up by the time the contract expires in December.

Say that gold rises by $30 to $630 in October. He decides to sell the contract to lock in a profit. The contract would be worth $63,000 and the profit would be $3,000. That $3,000 would be a profit of 150% in about two months.

Of course, gold could go down. If it had gone down $30 instead, then the contract would be worth $57,000 and Fred would realize a loss and a margin call would be made since he would have to kick in more margin.

Basic strategy #2: Going short

In going short, the speculator is making a bearish bet (thinking that the asset will go down in price) on that futures contract. The speculator gets into the contract by agreeing to sell and deliver the underlying asset at a set price. The bet is that the market price will fall below the set price and make a profit on the difference.

Let's say that after doing some research, Barbara expects the price of palladium to go down in the next few months. It is September and she decides to sell (go short) a January futures contract. Starting with an initial margin deposit of $3,000 Barbara would then sell a contract today, in September, at the higher price of $25,000, and buy it back after the price declined and before January. Remember, going short is a strategy to profit in a declining (or bear) market.

By November the price of palladium had gone down and the contract's market value had subsequently declined to $20,000. After buying back her short position, she nets a gain of $5,000 ($25,000 less $20,000). Of course, if palladium rallied (gone up) she would have lost money. Say that palladium's rally caused the futures contract's market value to rise and hit $31,000 in November. Barbara would have lost $6,000 (yes, the margin call would have been painful).

Basic strategy #3: Spreads

Going long and going short are out-and-out directional strategies. In other words you are 100% committed to a particular direction, either up or down. Once you get past those two, you get into strategies that offer some hedging opportunities for you the speculator. This means the next strategy . . . spreads.

Speculators can use spreads to take advantage of the price difference between two different contracts of the same underlying asset. Spread strategies are less risky than directional strategies which may make them good considerations for speculators with limited funds.

Check out the three common spreads in the following sections.

The calendar spread

In futures this is also called an intracommodity spread because it is done within the same commodity or asset. Say that Eddie is doing a calendar spread on silver futures. He will buy (go long) the further-out month while selling (going short) the closer month. Table 14-5 gives you an idea of how the trade looks.

Table 14-5		Calendar Spread Trade*			
Direction	*Asset*	*Price per ounce**	*Delivery month*	*Contract symbol*	*Contract value*
Going Long	Silver	13.675	Dec 08	SI Z8	68375
Going Short	Silver	13.42	July 08	SI N8	67100

Prices in this example are from NYMEX Web site as of 8/13/07.

As you can see, the contract for Dec '08 is valued at $68,375 and the July '08 contract is $67,100. The difference is $1,275. There is a difference because the market places greater value on the December contract.

In this spread, Eddie anticipates that the price difference between July and December silver futures will widen. Ideally, he'd like to see the short position (on July silver) go down and his long position (on December silver) go up. More likely the two contracts' prices will move in the same direction but Eddie is betting that the latter month contract will rise more so than the earlier month. He starts with initial margin of $600.

The spread between July silver (at $13.42 an ounce) and December silver (at $13.675) is 25.5 cents. Since a silver futures contract is on 5,000 ounces of silver, the total spread is then easy math (5,000 × 25.5 cents = $1,275).

As time passes and we get closer to July, Eddie is ready to cash out his calendar spread. Both prices rise but, as Eddie anticipated, the December contract increased more than the July contract. Presume that July silver rose by 20 cents and that December silver rose by 35 cents. After you do all the math, July silver is $13.62 and December silver is $14.025. The contracts are now worth $68,100 (July) and $70,125 (December). The spread has widened to $2,025. Eddie closes out the spread. The loss on the short position was offset by the greater gain on the long position. His profit is $750.

In the case that silver went down, the value of his position would go down but he could still realize either a smaller loss or a partial profit since his short position would become profitable. In the calendar spread, you have two positions that are like bets against each other. In a spread, you are hoping that your winning position is more successful than your losing position.

The inter-market spread

In the calendar spread, I provide an example to flesh out the concept to show you how a spread works. That gives you the gist so I won't have to beat you up with more detailed examples. You know now how a basic spread works. Beyond that, other spreads are the same basic structure but they may have some twist.

An inter-market spread is a spread with two contracts of the same month but two different markets. You go long in one market while you go short in another market. For example, you can go long one oil futures contract and short one natural gas futures contract.

The inter-exchange spread

This is another type of spread in which you go long on a contract in one futures exchange while you go short another contract in a different futures exchange. For example, the speculator may go long a contract on the Chicago Mercantile Exchange (CME) and go short a contract on the London Mercantile Exchange (LME).

And if you go long and short two dairy contracts then I believe it's called a cheese spread (but don't quote me).

Fundamental analysis

Fundamental analysis is usually used for stock investing but it is also done with futures. With stocks, fundamental analysis involves reviewing and analyzing the company's finances, operations, and its market to determine how well its stock will perform. As a side point it is recommended for long-term investing (measured in years).

The same methodology is used for futures but instead you are looking at the factors affecting its market price such as supply-and-demand data along with economic and political variables that could influence its value. As a side point, futures tend to be more short-term (usually measured in weeks or months). In general, futures certainly have a shorter time horizon than stocks.

Technical analysis

This method of analyzing futures and their underlying assets is based on the idea that market data, such as charts of price, volume, and open interest, can help predict the future (usually short-term) direction of market prices. The folks who use technical analysis believe that they can accurately predict the future price of a stock by looking at its historical prices as recorded in charts and accompanying data. In part, technical analysis makes the assumption (with the data) that market psychology influences trading in a way that enables predicting when futures prices will rise or fall. More on technical analysis in Chapter 18.

Fundamental versus technical

This one reminds me of that beer commercial where the two sides heatedly exchange those immortal words "tastes great!" and "less filling." That humorous debate made it sound as if it were one versus the other. But who says that you couldn't have both. I think that investing and speculating can taste great and be less filling. Well . . . you know what I mean.

In the financial world, fundamental versus technical analysis is a long-time debate but it's not a heated one. No fights break out in the bleachers and one group doesn't throw water balloons at the other one (at least not to my knowledge).

Considering my background and what has worked for me, I can be rightly called a fundamental analyst since I use an approach labeled value investing. Because stocks are predominantly a long-term pursuit, fundamental analysis works extremely well for me and others. For people with a long-term perspective, fundamental analysis generally wins out over technical analysis. As proof, in the historical pantheon of investors, you will find far more familiar names that are proponents of fundamental analysis and value investing (or some variation of these) versus those that use technical analysis and charting methods.

In the short term, it is a far different picture. Technical analysis has shown to be valuable and using it to determine short-term buy-and-sell trading decisions has shown some fruitful results. A great way to show these two approaches in action is to view an illustrative example. Here are the charts for silver for the year 2006 (Figure 14-1) and for the longer time frame 2000 to July 2007 (Figure 14-2). Combined, these two charts give a compelling perspective.

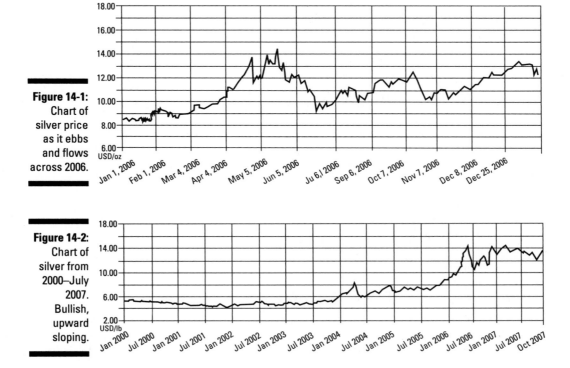

Figure 14-1: Chart of silver price as it ebbs and flows across 2006.

Figure 14-2: Chart of silver from 2000–July 2007. Bullish, upward sloping.

As you can see, silver was a trader's dream in 2006 (Figure 14-1): several solid rallies (up moves in the price) and several steep corrections (down moves in the price). Technical analysis was useful in determining conditions such as oversold and overbought which worked much like red flags to be careful and green lights which indicated a good entry point. During 2006, the funda-mentals were very strong for silver but that didn't stop the short-term price corrections that are really present in most bull markets (and certainly present in precious metals).

Chart B shows how the fundamentals could have been a superior catalyst to higher prices for silver. In the early part of the decade, silver was in the $4 to $5 range and it climbed to the $12 to $14 range during early 2007 (up 200%+, nearly tripling!). For long-term investors, fundamental analysis was more beneficial.

Chart A also shows you that the short term can be haphazard and irrational. The buying and selling sometimes doesn't make sense since market psychology — the hopes and fears that can drive the short-term buying and selling activity of speculators and investors — isn't always easy to figure out. This is why the short-term world of futures can be a risky and volatile one.

The bottom line is that both approaches have critical value to you since each approach's strengths can be used to your advantage. It is also why I tell my clients and students to put greater emphasis on the portion of their financial assets geared toward the longer term and a smaller portion (sometimes no portion!) geared toward the short term.

One of the biggest factors in the short-term volatility in futures is the very fact that futures contracts are paper assets. Since they are paper, you can create a lot of it and a lot of it can be bought and/or sold. Since relatively few contracts result in physical delivery, there is little standing in the way of creating more contracts than there are physical assets. Imagine that you have a market with one million cars (the underlying asset); what if there were two million car titles being bought and sold? If too many car titles are being sold (going short), then that would temporarily and unusually drive the price down. If too many car titles are sold, then it would have the same effect upward. (The CFTC is there to prevent excessive events such as these.) Too much paper can warp the market price causing it to go unnaturally higher or lower as the case may be. The bottom line is that futures can cause more radical swings in the price in the short-term that you should be aware of.

Futures versus Options on Futures

By now, you are aware of the good, the bad, and the ugly of futures. It is a great market but it can be a scary one as well. You've seen the statistics about how 80% of all speculators lose money in futures and that some of them end up in insane asylums singing Ethel Merman tunes while drinking Bosco without milk (it's tragic). Darn it . . . I don't want you to end up like that. So let's add some sanity to this topic.

Futures trading is great but it's obviously not for everyone. I think that if beginners and those with limited funds want to experience the action of the futures market, let me give you a strong recommendation to make your entry into this market a little less stressing. Consider long-dated options on futures.

As I have written earlier in this chapter, options on futures can have some of the same characteristics as futures. Generally, options are considered a short-term market much like the futures market. But notice that I write long-dated options. This is a reference to a category of options referred to as LEAPs.

LEAP stands for Long-term Equity AnticiPation securities. They are almost exactly like regular options but with one small yet major difference: LEAPs have a much longer time frame or shelf life. Typical options can be up to nine months before expiring; LEAPs can go out three years (and sometimes longer).

Take the above example on silver. I use silver futures contracts with deliver dates in 2008. But there are options on other silver futures contracts that go out to December 2009 and even into 2010 and 2011. You would have to speak with your broker about availability since sometimes the exchange may issue futures contracts and options for far-away dates but there may not be an active market yet for those securities. Granted, futures may have more profit potential, but to me the difference in profit potential is not that attractive since options on futures has very significant profit potential (with significantly less downside potential).

The bottom line is that long-dated options on futures (LEAPs) are a good way to play the futures market and still have the benefit of more time so that you can work the market as if it were an investment (albeit an aggressive one).

Futures Resources

I hope you took note of all the specific futures-related Web sites I list in this chapter. Most of them have great educational materials, resources, and links that will be invaluable to you (or is that valuable?). Check out these futures-related Web sites that I think have merit for your research:

- Futures Source (www.FuturesSource.com)
- Futures Buzz (www.FuturesBuzz.com)
- Futures Spot (www.FuturesSpot.com)
- Commodities Links (www.commoditieslinks.com)
- Futures Knowledge (www.FuturesKnowledge.com)

Some general sites that have great information on the basics of futures, including explanations for beginners on futures terminology:

- Investor Words (www.investorwords.com)
- Investopedia (www.investopedia.com)

Chapter 15

Options

In This Chapter

▶ Understanding how options work

▶ Checking out conservative, defensive, and aggressive options strategies

▶ Taking a look at options on metals and related investments

*O*ptions are a very versatile investing vehicle. They are, in fact, excellent adjuncts to your precious metals strategies. When I do my national seminars on options it does my heart good to see bright-eyed and bushy-tailed budding investors in the room have their faces light up when I discuss the possibilities in their portfolios with options.

In general, options are extraordinarily versatile. They could be used for short-term or long-term purposes. Options can be used for games or for generating income. You can use options for speculative purposes or for defensive or conservative purposes. Options can be used for that home run shot where you can make a lot of money very quickly or more to protect the position in your portfolio. You can use options to make money in up markets or in down markets. Heck, you could even use options to make money on flat or boring markets. In my eyes, options are perfect for virtually any portfolio depending on the type of strategy and the suitability.

Options are perfect vehicles for the portfolio of any precious metals investor no matter how aggressive or conservative. Options can help juice up the returns on your precious metals stocks or futures contracts, or to get help removing some of the risk and volatility that are part and parcel of the world of precious metals.

How Options Work

An option is a contract. That means that it doesn't have its own value; it derives its value from the underlying asset.

There are two types of options contracts: Calls and puts. A call gives the holder or buyer of the options contract the right but not the obligation to buy the underlying asset at a specific price during the life of the options contract. The most important risk to be aware of when you buy an option, whether it's a call or put, is that it has a finite life and it can, if you're not careful, expire worthless. An example of a call option is presented in the next section.

Please keep in mind that you can get an option on virtually any marketable securities that are assets available in the marketplace today. You'd be surprised what things have options on them to be a stock exchange-traded funds or futures contract (or other assets). There are two parties in this contract, the buyer of the contract and the seller or writer of the contract.

The call option

A call option is a contract that allows the buyer to buy a particular security or asset at a specific price on or before a certain date. Just like the name implies it is an option; the person doesn't have to buy the underlying asset but can if she wants to. Buying a call option is basically a bet that the underlying asset will be going up. The call buyer is also betting that the up move will occur before the option expires. It may sound a little complicated or legalistic, but when you think about it, you have probably bought options in the past. An option is a contract that is called a derivative. A derivative is nothing more than an investment vehicle that derives its value from something else; in the case of options, it is the underlying asset.

Examples of options

A good example of a call option is, believe it or not, a lottery ticket. A lottery ticket doesn't have its own value; after all, it is a small piece of paper with printing on it that doesn't have any intrinsic value. So why would you buy a lottery ticket? A lottery ticket is like a call option on a set of numbers. A lottery ticket derives its value from of that particular set of numbers. It is also like options because it has a finite shelf life; that lottery ticket will expire on the specific date whether or not those numbers come out. If those lottery numbers do not come out then a lottery ticket is worth nothing. The lottery ticket is worth nothing because it derives its value from numbers that are worthless. Now how much would that lottery ticket be worth if the numbers did come out? Obviously that lottery ticket would, at that point, be worth a lot of money, maybe even millions. That lottery ticket would then derive its value from some very valuable numbers. Of course, if the winning numbers come out after the date of a lottery ticket, then that ticket will expire worthless.

Another example of a derivative is a car title. A car title has very little of its own intrinsic value; it's just a piece of paper with some printing on it. The car title derives its value from the underlying asset, which is an automobile. Now if that automobile is a brand-new luxury car with all the bells and whistles then that car title is a very valuable piece of paper. Since the underlying asset (a snazzy car) is very valuable, therefore that piece of paper is valuable. Now if that car title is attached to a 1987 Ford Torino that was in several major car crashes and had a dozen paint jobs then that car title would derive . . . uh . . . much less value. But let's get back to the call option.

In a call option there is a call buyer and the call writer (the seller). The call buyer pays money (called the premium) to the call writer for the right but not the obligation to buy the underlying asset at a specific price before the call option expires.

When you buy an option, whether it is a call or a put, it has a finite life. The remaining life may be a few weeks or a few months or even a few years but an option will expire sooner or later. The great danger with buying an option is that you could easily lose 100% of your money in the option if you're not careful. For this reason I do not call an option an investment. Buying options is not investing; it is really speculating. Speculating is a form of financial gambling and is much different than investing. Investing means you're putting your money into an asset (such as stock or real estate) that has changeable value, is appropriate for your situation, and can be held for long-term. This asset may fluctuate but if chosen wisely will keep trending upward. An option, on the other hand, is a wasting asset; as the clock keeps ticking the options time value will keep shrinking until it expires.

All the possible outcomes for the option buyer

The following are all the possible outcomes for an option buyer:

- ✔ **The option can expire worthless.** This is the worst case. Since options can be relatively inexpensive (versus buying the underlying investment outright) it's not a huge loss.

- ✔ **You can exercise the option.** Exercising means that you formally request the buying (or selling) of the underlying security at the strike price.

- ✔ **The option goes up a lot.** You can cash out the option and make a large profit.

- ✔ **The option goes up a little.** The asset is moving in your direction and you cash out the option and still make a profit.

- ✔ **The option goes down.** You can still cash out the option and take a loss and at least recoup some of your investment.

All the possible outcomes for the option writer

The following are all the possible outcomes for an option buyer:

- ✔ **The option can expire worthless.** You would like this outcome since it means that you get to keep 100% of the option premium.

✔ **You may have to fulfill your obligation.** The buyer may exercise the option, forcing you to make good on your obligation to buy or sell the underlying security.

✔ **The option gains value.** You can let the option trigger your obligation or you can buy back the option at a loss.

✔ **The option loses value.** You can let the obligation keep losing money or you can buy it back at a profit and remove your obligation.

✔ **You want to get out of your obligation regardless.** You don't want the risk of the option being exercised by the buyer so you buy back the option at a loss (or gain) depending on the price of the option.

The strike price

The strike price is the agreed-upon price in the call options contract. Say that you have an option on a stock called Juggernaut, Inc. Presume that the stock is at $50 per share and that the strike price in the option is $55. Say that the option will expire in nine months (for our example, assume that the expiration date is December 31, 2007) and the option buyer paid $200 for this option. The option buyer is making a bet that Juggernaut will rise toward $55 (and hopefully exceed it) before the expiration date. In this case, the stock's market price ($50 per share) is $5 away from the strike price. I like to remember it as the market price striking the target price, in this case $55. If the stock stays at $50, or rises a little to $51 or $52, or if it falls lower, then the odds are that the option will expire worthless. After all, what good is an option that gives you the ability to buy the underlying asset at $55 if the market price is much lower? As you can see, the relationship of the asset's market price to the option's strike price is critical.

Is it out, in, or at the money?

In the world of options, the relationship of the asset's market price to the option's strike price is referred to as being

✔ **Out of the money:** With Juggernaut at $50 and the strike price at $55, it is out of the money (or OTM). OTM options don't have much value.

✔ **At the money:** If Juggernaut rose to $55, it would be called at the money (or ATM) because the market price of the asset and the strike are basically equal. ATM options are usually more valuable than OTM options.

✔ **In the money:** If Juggernaut rose to, say, $60 per share, then the option would be referred to as being in the money (or ITM). The ability to buy a $60 stock for only $55 (the option's strike price) makes that a valuable option. An ITM option would have both time and equity value.

Time and equity value

So what does the whole buyer get with the premium that he or she pays? The call buyer gets some combination of time and equity value for the premium is paid. This is what you get, given the three different types of options:

✔ **Out of the money (OTM).** The premium you pay for an OTM option is 100% time value. There is no equity value. You have paid for the duration of the option's life span. An OTM is the riskiest option because basically you're hoping the underlying asset will hit the strike before it expires. OTM options are the cheapest options but also the riskiest. Fortunately, the full amount at risk is only the premium you paid for the option, which is probably not that much. Because with an OTM option you are buying time, the longer the option period, the more time you are paying for. A one-year option, for example, is more valuable than a six-month option. In a nutshell, an OTM option is 100% time value and 0% equity value.

✔ **At the money (ATM).** Because the market price of the asset and the option's strike price are equal or at parity, the value of the option is definitely higher than the OTM (all things being equal). An ATM option has a greater chance of becoming an in the money option, of course, than the OTM option. The value of the ATM is still more time value than equity value but at least it has some equity value. How much equity value it has will depend on the marketplace.

✔ **In the money (ITM).** Because the market price of the asset is higher than the option's strike price, the value of the option is higher than either the ATM or the OTM. The OTM option will have plenty of equity value and will still have some time value, of course, depending on how much time is left.

Keep in mind that the time premium is the amount buyers are willing to pay for the options contract above its intrinsic value on the chance that, at some time prior to its expiration, it will move into the money. Out-of-the-money options all carry time premium since their intrinsic value is zero, as is that of at-the-money options.

The time premium for the in-the-money options contract is the amount that exceeds the option's intrinsic value and reflects the possibility that the options contract may move deeper into-the-money. The time value of an options contract shrinks as the expiration date approaches, with less and less time for a major change in market opinion, and a decreasing likelihood that the options contract will increase in value (see Table 15-1).

Table 15-1	**Calls: Options Rights and Obligations**
Buyer	*Seller (the writer)*
Pays premium	Receives premium
Has the right to buy a futures contract at a predetermined price on or before a defined date	Grants right to buyer, so has obligation to sell futures at a predetermined price at buyer's sole option
Expectation: Rising prices	Expectation: Neutral or falling prices

The put option

The put option is making a bet that the underlying asset is going down in price (see Table 15-2). Officially, a put option is a contract that gives the buyer (or holder) the right to sell a certain quantity (such as 100 shares of stock) of an underlying asset (such as stock or a commodity) to the writer (or seller) of the options contract at a specified price (the strike price) on any business day during the life of the option.

Table 15-2	Puts
Buyer	*Seller (the writer)*
Pays premium	Receives premium
Has the right to sell a futures contract at a predetermined price on or before a defined date	Grants right to buyer, so has obligation to buy futures at a predetermined price at buyer's sole option
Expectation: Falling prices	Expectation: Neutral or rising prices

Resources for beginners

The best places for information on options for beginners are the options exchanges and their trade association. I have visited their Web sites and there is a tremendous amount of free educational materials, tutorials, publications, and so on. And for good reason . . . the exchanges make money each time an option is transacted. The more people doing more options transactions makes for a happy exchange. Given this, you might as well take advantage of the great stuff they offer.

- ✔ Chicago Board Options Exchange (www.cboe.com): This is the place to turn to for extensive information regarding options on stocks, stock indexes, and exchange-traded funds.
- ✔ New York Mercantile Exchange (www.nymex.com): Offers extensive information on futures and options on futures.
- ✔ Options Industry Council (www.888options.com): Lots of publications as well as free audio and video online programs well suited for beginners.

Working Out Your Options

If you want to go forward with options, you will need to open an account first. The account will need to be funded and approved for options.

For options on stocks, stock indexes, and ETFs, a standard stock brokerage account will suffice. For options on commodities/futures, you will need a futures brokerage account.

Understanding the orders

When you are ready to make your options order, the brokerage firm has information, customer service, and so on to help you. Just keep in mind the following order types to make sure you are transacting what you truly want:

- ✔ **Buy to open:** When you initially buy an option (put or call), you are essentially paying for a contract that you are opening. This is called buying to open.

- ✔ **Sell to open:** When you are writing an option (put or call), you will receive income (as in selling) for opening a contract. This is called selling to open.

- ✔ **Buy to close:** When you want to end or close out an option that you had initially sold then you need to buy back your option to close it out. This is called buying to close. This completes the option described in #2 above.

- ✔ **Sell to close:** When you want to cash in an option that you had initially bought, then you need to close out the options contract. This is called selling to close. This completes the option described in #1 above.

Something for Everyone

The great thing about options is not only their tremendous versatility and profit potential but also their ability to generate income and even act as insurance. In this section there are important points for investors and speculators and even worrywarts.

For those seeking gains

If you're going to speculate, there's probably no better way than with options. With unlimited upside and limited downside then you have some powerful speculative advantages. I remember buying some options on silver futures in 2005 that I bought for only $600 when silver was $7 an ounce. Within six months, silver went over $10 an ounce and each of the silver futures options was worth $10,000! This is the power of leverage.

As I mention, options are technically not an investment. They have a limited life and you must do something with them before they expire. It's hard to invest with something that typically has a life span of only nine months.

That's why I like to invest using LEAPs. LEAPs are Long Term Equity AnticiPation Securities. They are the same in almost every way to regular, conventional options with one basic difference: the time span. LEAPs can be up to three years.

Income strategy #1: Writing covered calls

When you write a covered call option you are doing so on a stock that you own. For every 100 shares of an optionable stock that you own, you can write (sell) a covered call and you will receive income (the premium) in exchange for an obligation to sell your stock at the strike price.

Say you own 200 shares of General Electric (GE). On July 17, 2007, the stock price was at $40.71. On that day you could have written a September 2007 call with a strike price of $42.50 and received income of $60 (leaving out the brokerage commission to keep it simple). Since a call contract is tied to a round lot of 100 shares, you could write two calls (since you own 200 shares of GE) and receive income of $120. In return, you take on the obligation to sell your GE stock for $42.50 on any business day starting July 17 up to the third Friday in September 2007 (September 21).

It is a risk-free transaction for you in that there is no risk of financial loss. If GE rises significantly between now and September '07, the worst that happens is that you sell your stock for $42.50, which is a higher stock price than the $40.71 price back on July 17. In addition, you made the $120 income from the premium.

If GE stays flat or goes down during that time frame, you get to keep the stock and the $120 from writing the two calls. Owning the GE stock has the usual risk of stock ownership; it can go up, down, or stay at the same level. Writing the covered call itself was a risk-free way to generate income from your stock holding.

Covered call-writing is a safe strategy because you own the underlying stock and you simply relinquish the stock if it hits the strike price during the life of the option. In others the call obligation is covered. However, it is possible to write an uncovered (or naked) call. This transaction is the same in every way to the covered call except in one crucial way: you don't own the stock. You could still generate the $120 (in this example) but if GE rises and hits $42.50, you have the obligation to provide the 200 shares. Since you didn't own the stock, you'd have to buy it at the market price (whatever price it is) and then sell the stock for the $42.50. If GE rose to $50 a share, you'd end up losing $7.50 per share (on a 200-share transaction, you'd lose $150). In a naked call you gamble to make the option income but you risk having a loss that is potentially far higher. Most retail investors are not allowed to do naked call-writing because of the potential loss involved.

Income strategy #2: Writing puts

If you write a put, you will receive the premium income but in exchange you take on the obligation of buying stock if it hits the strike price. To be allowed to write a put in a stock brokerage account, you would need to have enough cash (or marginable securities) in your account to handle the potential purchase. This is referred to as a cash-secured put. If you are using marginable securities, that is referred to as a portfolio-secured put. So-called naked puts — writing puts without sufficient cash or marginable securities in the account — are not allowed for most retail investors.

You should only write puts on stocks that you personally would enjoy owning. Writing a put option on a mining stock when it's at $35 and the put option's strike price is at $30 means that (in the worst case) you end up buying that stock at the bargain $30 price. The great thing is that the premium helped make the purchase more affordable. If you got $100 for writing the put and you need to buy the stock, you really only had to shell out $2,900 for those 100 shares ($3,000 total purchase price less $100 premium).

Minimizing Risks with Options

In earlier chapters (such as Chapter 11) I briefly discuss the defensive merits of stop-loss orders and the active strategy of trailing stops. (I discuss these in great detail in *Stock Investing For Dummies*.) But a great way to protect your stock position without having to sell the stock is to buy a put on your own stock. In this instance, the put that you bought would be referred to as a protective put. It becomes a form of insurance.

Say you own 100 shares of Gold Corp. (GG). On July 17, 2007, GG stock was at $25.85 per share. Let's say that you are nervous about market conditions for GG during the summer but you don't want to sell the stock because you bought it a few years earlier for only $8 a share and you don't want to generate a capital gain and worry about the tax on the potentially large gain (the market price less the $8 original cost). Instead of selling, you could buy a put for that time period to give you some protection. You could buy an out-of-the-money put with a strike price of $22.50 that expires in October '07 for only $60 or an at-the-money $25 put (same month) for $140, depending on your expectation or budget. The bottom line is that you could buy insurance in the form of a protective put for very little money.

Some Profitable Combinations

Options are very versatile and once you figure out the ins and outs of a call option and a put option, then consider moving to the next level, combinations. There are literally dozens of options combinations. You can structure combinations to give you some exceptional profit potential.

The zero-cost collar

This is one of my favorite combinations and it is an ideal strategy for those who are worried about some positions in their portfolios. It is done on a stock that you own and hold in your stock brokerage account.

It is the act of buying a put option and selling a call option on your stock. Let's say that you have 100 shares of Kabluki Corp. (KC) stock and it's $30 a share. Say the market is getting you concerned about your KC stock. Table 15-3 shows you the zero-cost collar for it.

Table 15-3	Zero-Cost Collar in Action		
Buy/sell action and option details	**Type**	**Comment**	**Cost**
Buy 1 KC Put. Strike price $27.50. Expires 10/19/07	Out of the money (OTM)	This is a protective put (you worry about your stock going down)	$100 (you are paying premium)
Sell 1 KC Call. Strike price $32.50. Expires 10/19/07	Out of the money (OTM)	Bearish (you expect KC stock to go up)	$100 (you are receiving premium)
			Total cost = $0

Take a look at some of the advantages. First of all, the zero-cost collar is protection that can be structured to cost nothing (I like that part). If KC stays flat or trades sideways, the options could expire. But . . . you don't mind because the collar was zero-cost.

What happens if your concern is realized? Although your KC stock would fall, the zero-cost collar helps offset the loss with a double profit. When KC falls, the put would increase in value and the call would lose value. The put would be profitable. Now about that call, keep in mind that you didn't buy it; you sold it. Therefore, you don't mind it losing value. Now you can buy it back at a profit.

The drawback for the zero-cost collar is if the stock goes up significantly during the options period. Because part of the combination was a covered call, that limited your upside since you would have to end up selling the stock at the strike price of $32.50.

The straddle

This is a simple combination. It is buying both an at-the-money call and an at-the-money put (both have the same expiration date) on the same security or asset. Table 15-4 uses the example of XYZ Mining Corp. stock (price: $50/share) to illustrate the straddle.

Table 15-4	The Straddle in Action		
Buy/sell action and option details	**Type**	**Profit direction**	**Cost**
Buy 1 XYZ Call. Strike price $50. Expires 12/21/08	At the money (ATM)	Bullish (you expect XYZ stock to go up)	$300
Buy 1 XYZ Put. Strike price $50. Expires 12/21/08	At the money (ATM)	Bearish (you expect XYZ stock to go up)	$300
			Total cost = $600

The basic point behind the straddle is very simple. You don't care where XYZ goes (up or down); you just want the stock's price to go somewhere and to get there very fast. Certainly you want it to happen long before the call and put options expire. Options strategies can be bullish, bearish, or . . . in this case . . . neutral. The buyer of the straddle combination is neutral and is actually seeking high volatility, a big move in either direction.

The bottom line is that you're hoping (in this example) that XYZ goes far enough (up or down) that one of the options in the straddle will be worth more than the two options combined. If, for example, XYZ skyrockets to $70 then the call will be very valuable while the put would be almost worthless. In the case of a $70 stock price, the call option would be worth at least $2,000 while the put option's value would be close to zero. In valuing the straddle, we can say that it went from a total value of $600 to an ending value of $2,000 (the $2,000 call plus the $0 put).

Keep in mind that technically the option combination that I just described is the long straddle. It is labeled long because you bought both options. Yes . . . there is a short straddle but that goes beyond the scope of this chapter. Any and all the options terms, combinations, and so on are described more fully through the resources referenced.

Options in the World of Precious Metals

The first half of this chapter goes into the generic world of options. That is necessary so that this section has more meaning for you. There are many ways to play precious metals, and few vehicles give you as much oomph as options and when you combine precious metals with options you get . . . uh . . . ultra-oomph!

Options on mining stocks

The most obvious place to start is with the mining stocks, of course. There are many mining stocks with options available on them. Some of the stocks have a limited number of options issued and available for investors (many may have only 30 to 40 options going out nine months at most). Some of the larger and more active companies have 75 to 150 (or more) options available and stretching out two years or more (LEAPs).

Here are some examples of mining stocks with plenty of long-term options on them:

As of July 2007, the gold-mining company Gold Corp. (GG) had options available all the way out to January 2010 (a full two and a half years of time value). By the time you read these words, there will be options on that stock going out to 2011 or beyond.

As of July 2007, the silver mining company Pan American Silver (PAAS) also has options stretching out to January 2010.

As of July 2007, Cameco (CCJ), the world's largest uranium miner, had options available with expiration dates of January 2009.

These are mentioned as illustrative examples of what you can find. To find more, use the resources in Chapter 12 (on mining stocks) to locate and analyze the mining stocks. The same resources can tell you if they have publiclytraded options available. If the answer is yes then you can go to the Chicago Board Options Exchange (www.cboe.com) and do your search. At the Web site you can simply enter the stock symbol of your choice and up will come the options available for you to buy (or write).

An option with no expiration?

As you know by now, options are a derivative. But some stocks, especially those in the natural resource sector, can be considered derivatives of the resources they are involved with. Last year I bought a stock that had substantial nickel reserves at about the time that industry reports mentioned supply problems with . . . you guessed it . . . nickel.

That stock quickly shot up from under $2 a share to over $5 in a few weeks for a solid gain of over 150%. That was a small company and it had no publicly traded options. However, the stock acted like an option on the underlying asset. In that case, $200 would have gotten you 100 shares and the best part is that, unlike options, there is no expiration date.

Options on ETFs and indexes

Exchange-traded funds (ETFs) are explained in detail in Chapter 13 but I bring them up here because they are great underlying assets for options investors and speculators. Why make a bet on an individual security when you can bet on an entire industry or sector? You have the same options capabilities as with stocks. You can buy or sell options (calls or puts).

ETFs with options

The following are ETFs with options available (go to www.cboe.com for details):

PowerShares DB Base Metals Fund (DBB)

PowerShares DB Precious Metals Fund (DBP)

PowerShares DB Silver Fund (DBS)

PowerShares DB Gold Fund (DGL)

PowerShares DB Commodity Index Tracking Fund (DBC)

Indexes with options

There are indexes that have options available on them. A good example of one is found at the Chicago Board Options Exchange: the CBOE Gold Index (GOX). It is an equal-dollar weighted index composed of 10 companies involved primarily in gold mining and production.

Options on futures

For those who want the speculative, high-flying, ultra-aggressive, no guts, no glory home-run or strike-out type of vehicle, then you may want to consider options on futures. Check out Table 15-5 for a list of options on regular futures.

Table 15-5	Options on Regular Futures		
Type of asset	*Size of standard futures contract*	*Trading symbol*	*Comments*
Gold	100 troy ounces of gold	GC	Precious metal
Silver	5,000 troy ounces of silver	SI	Precious metal
Platinum	50 troy ounces of platinum	PL	Precious metal
Palladium	100 troy ounces of palladium	PA	Precious metal
Uranium	250 lbs. of U308	UX	Metallic mineral used for fuel
Aluminum	44,000 lbs. of aluminum	AL	Base metal
Copper	25,000 lbs. of copper	HG	Base metal
Zinc (London SHG)	10 metric tons	LZ	Base metal

Table 15-6 gives you a list of options on mini-futures.

Table 15-6	Options on Mini-Futures		
Type of asset	**Size of standard fcontract**	**Trading symbol**	**Comments**
Gold	50 troy ounces of gold (NYMEX miNY (tm) Futures)	QO	Traded on NYMEX
Gold	50 troy ounces of gold (Chicago Board of Trade mini-sized futures)	YG	Traded on CBOT
Silver	2,500 troy ounces of silver (NYMEX miNY (tm) Futures)	QI	Traded on NYMEX
Silver	1,000 troy ounces of silver (Chicago Board of Trade mini-sized futures)	YI	Traded on CBOT
Copper	12,500 lbs. of copper (NYMEX miNY (tm) Futures)	QC	Traded on NYMEX

To do options on futures contracts you will need to open a futures account with a commodities brokerage firm.

Golden Rules for Options Success

After watching this stuff for a few decades, teaching thousands of students, and actively doing this day-in and day-out, you get to see what works and what doesn't work. Options are no different and there are points you can take to heart and apply to your own approach. After reading everything in this chapter (along with the referenced resources), here are my golden rules for options success:

✔ **Understand the risk in your strategy.** One speculator (I'll call her Betty) whom I know of got a hot tip about Google and she immediately bought 350 call options. Google didn't go the way she expected and she lost over $20,000. Yikes! The options she bought had only two month of time! That wasn't speculating . . . that was outright gambling. She might as well have taken the money to the local casino. Betty didn't realize the risk of too little time. Also, why risk so much and so quickly?

Sometimes the risk is not only in the vehicle (very short-term options) but also in the person. Impatience and acting without full information can be the kiss of death for your wealth-building aspirations.

✔ **Decide what is the most amount you will put at risk.** In the previous example, Betty sadly told me afterward that the total amount she lost was a huge portion of her financial assets. It caused her hardship. As I have said before, buying options is a speculative pursuit and you should know from the start how much you are willing to risk and no more. You should ask yourself "Will my lifestyle change drastically if I lose this money?" If that question makes you pause then you may be risking too great a portion of your financial assets. For most people, 5% to 10% of your financial assets is a comfortable limit. As you grow in experience, confidence, and skill, you can certainly do more.

Options buyers should get the longest time available and a strike price that is as close as possible.

What do you think has less risk — a two-month option or a two-year option? The more time, the better for you. What has a greater chance to be profitable — an at-the-money option or an out-of-the-money? The closer the strike price is to the market price, the better your chance of success. However, there is a better strategy: both! Your chance for success grows tremendously in your favor when the option you have has lots of time value and a strike price that is as close as possible to the market price.

✔ **Options sellers should write options that are out-of-the-money and expire soon.** For those who are seeking income and want to minimize the risk of covered call-writing (the chance that your asset could rise to hit the strike price and subsequently trigger the sale of your asset) then you need to do the mirror opposite of the call buyers' golden rule (see previous paragraph). In covered call-writing you need to write options that are comfortably out-of-the-money and will expire in a relatively short period of time.

✔ **Don't wait too long to cut your losses.** Depending on the type of asset and the length of the option period, the options buyer should not wait until the final 30-60 days of the option's remaining time value. Again, options are wasting assets and they have the tendency to lose value dramatically in the final weeks. If you cash out the option while it still has some time value then you can recoup some money and put the funds to work elsewhere.

✔ **Take profits sooner if you've written options.** Say that you wrote a covered call and you got some nice premium. If that stock had made a sudden yet temporary drop in price, consider taking some profits by buying back your covered call. Why wait? Make a profit and now your stock is free and clear and available for the next covered call option.

✔ **Focus on one area and become an expert in that area.** Ironically, the best advice I could give for options success actually is not even an outright options-related recommendation. Remember the point earlier in the chapter that options are derivatives. Options derive their value from the underlying asset so it behooves you to become as knowledgeable and as proficient as possible in that underlying asset. Nine times out of ten, when you see someone who consistently succeeds with options it is usually due to the fact that the investor or speculator focuses on that singular area and gets really good at understanding that particular market.

✔ **Be patient.** Maybe it's a sign of the times and how people's behavior and attitudes have changed. Investors today are way too impatient and they are not allowing their strategies and investment choices enough time to bear fruit. Frankly, it has me scratching my head. It makes me compare today's investors with investors from the good ol' days. When I first started my financial planning business (1981 — I hope I don't look that old) investors had a more realistic attitude toward time and money. Back then long-term was over five years, intermediate term was one-to-five years, and short-term was one year or less. Today, I think that investors measure long-term in months, intermediate term in weeks, and short-term in days. I asked one client what his idea of short-term was and he responded, "What time is it?"

Long-term investors and speculators have an easier time making money. When I say "long-term" that doesn't necessarily mean you have wait 57 years for something good to happen. It doesn't even have to mean a rigid time frame. It means giving the market *enough time to discover what you have*. One client I had bought a uranium stock in 2005. It went up nicely that year but then during 2006 it stayed in a narrow trading range of $35-$39 for many months. His impatience resulted in the sale of that stock but guess what? Within a few months that stock shot up to $56 a share for a gain of 43% in less than five months. The call options on that stock did even better. Over the years I have seen many situations where impatience led some to prematurely sell (or buy) something, needlessly resulting in a loss or missed opportunities.

Options Resources

You can find lots of great educational resources on options at the library and on the Internet. Here are some that you shouldn't miss:

✔ Educational resources:

- There are free and/or low-cost seminars on options. The Options Industry Council conducts free options seminars across the country. You can get on their mailing list by calling them at 888-OPTIONS or visiting their Web site at `www.888options.com`. I, the author of this book, also conduct seminars and you can find out more through the Web site `www.PaulMladjenovic.com`.

- Many of the general financial Web sites such as `www.invest opedia.com` and `www.invest-faq.com` have great how-to articles and tutorials.

- The options exchanges offer some excellent information on options and you can visit their Web sites at `www.cboe.com`, `www.nymex.com`, `www.cbot.com`, and `www.cme.com`.

✔ Publications for options on stocks and futures:

- SFO Magazine (*Stocks, Futures & Options Magazine*) (`www.sfomag.com`)

- *Futures Magazine* (`www.futuresmag.com`)

✔ Internet resources for options on futures:

- Commodities Futures Trading Commission (`www.cftc.gov`)

- Future Source (`www.FutureSource.com`)

- Future Spot (`www.futuresspot.com`)

Part IV:
Investment Strategies

The 5th Wave   By Rich Tennant

"This is interesting. China is currently the world's largest consumer of iron, iron ore, and irony."

In this part . . .

Are you short-term or long-term in your investing or speculating focus? Are you a trader? To move forward you need to get familiar with the tools, brokers, and strategies available. In addition, check out this part to find out how to keep more of your money after all is said and done (that's fancy phrasing for — taxes).

Chapter 16

Choosing a Trading Approach

- -

In This Chapter

▶ Preparing to trade

▶ Deciding on a trading vehicle

▶ Taking a look at your outlook as well as trading strategies

- -

Now for the fast-moving stuff. No one can say that trading is boring. It's exciting — you just want any discussion on trading to remove the words "losing," "terrifying," and "depressing." Of course, trading can be profitable. Some have successfully turned trading into a full-time pursuit, which takes lots of preparation and the right frame of mind because it can be stressful at times. For a trader, "risk" and "volatility" aren't just words — they're a way of life.

Before you delve fully into the trading lifestyle, prepare yourself by first going through this chapter. In this chapter, I explain the various ways to prepare before you even begin trading as well as provide you with trading techniques to choose from. Preparing yourself and choosing the technique that best fits you should make the exciting life of trading more profitable and, well, exciting.

Being a Boy Scout — Being Prepared

You get the greatest benefit from trading before you risk a single copper cent (well, it was copper but now it is an even baser metal) by simply preparing. Preparing for your trading pursuits puts you one step closer to successful trading. The sections that follow give you an idea of how to begin thorough preparations.

Know the difference between trading and investing. You're *trading*, which is a short-term play on the price moves of the asset. Whatever you're putting your money in you will be exiting this position relatively quickly. You just want someone to buy your stuff at a greater price and sooner rather than later. *Investing* means you're putting your money into an asset that has value today and that will essentially have long-term appreciation. This difference dictates your approach, and it dictates which trading philosophy will dominate your actions: fundamental or technical analysis. For traders, "price action" is the name of the game.

Be a voracious reader

Paraphrasing the Bible, there is nothing new under the sun, and this is true in the world of trading. The technology may have changed radically, but the principles have remained the same. The successes and failures and strategies are there to learn from. The same way that investors should read the works of Benjamin Graham and Warren Buffet, traders should be reading about the exploits, strategies, and principles of great traders such as Jesse Livermore.

Two books worth checking out are *Reminiscences of a Stock Operator* by Edwin Lefevre (Wiley) and *How to Trade in Stocks,* which is in Jesse Livermore's own words but re-published by McGraw-Hill (the copyright is owned by Richard Smitten). Livermore was a very colorful character and he became the ultimate rags-to-riches trader as he made and lost four fortunes during the 1920s and '30s before he committed suicide in 1940. The amazing thing is that he started trading as a teenager starting with five dollars (of course, five dollars was worth something back then), and he went on to make millions.

In today's market, some of the well-known traders include long-time veterans such as Larry Williams, Dick Diamond, and Roger Wiegand. There are more trading resources at the end of the chapter.

Have your plan

Making a trade with guesswork or your best hunch isn't enough. Certainly not if you want to build wealth over time. You need to decide on a "framework" for your trading activity. Your plan needs to have clear and detailed answers to questions such as

- ✔ How much risk capital will you be playing the trading game with?

- ✔ What is your outlook on the asset in question (bullish, bearish, or neutral)?

- ✔ Will you be focused in a single specialty or diversified in different assets (such as stocks, options, futures, or in different industries or commodities)?

- ✔ At what point will you enter a trade? What signals will you use (such as technical indicators explained in Chapter 18) as your entry points?

- ✔ How long will you stay in your position? Will it be a fixed time period or until a particular event occurs (such as when it hits a certain price)?

- Will you be doing any hedging (a way to reduce risk by having positions in your account that go up if your main positions go down — more on this later)? If so, what kind of hedge?

- What will you do if the position goes down in price during your time period? Buy more or get out?

- If you buy more at the lower price, what will you do if the asset's price goes down even further? What price level or loss percentage will you tolerate before you decide to cut losses?

- At what point do you cash out profitable trades? Is there a specific amount or is it based on market events (such as technical indicators or news from the industry)? Will your advisory service or software tell you?

- At what point will you say "I can't take it anymore! I'm getting a job!"

You know what I mean. There are questions that no amount of research or professional guidance can answer for you. Much of what you do will be your own preferences and trading objectives. The point is to address these issues since the market will make mincemeat of traders who are throwing darts.

Decide your market

Until you are proficient and experienced with markets in general, your best bet is to get thoroughly familiar with a particular asset. Trading gold and silver futures is fine, but beginners shouldn't trade gold and silver along with grains, pork bellies, and natural gas. Commodities can be specialized and the more you stretch yourself over more different markets the less time you will be able to become an expert in the markets that count.

Many of my clients and students will get extremely knowledgeable about a particular market and then apply their trading strategies to take advantage of what they researched and studied. I know one trader who researched the oil market and turned $90,000 into $1.4 million in about 18 months during 2004–2005. He used bullish trading strategies such as buying call options that skyrocketed as oil prices nearly doubled during that time period. He did focused research (specialized in one market) and focused activity (buying call options), which paid off handsomely.

Specializing in one market will make it much easier to succeed. After the learning curve to find out what's going on then you can monitor the market and implement your trades in minutes.

Fortunately, precious metals investments have a lot to offer. You can specialize in different metals (precious or base) and in different vehicles (stocks, futures, and/or options). What will you concentrate on?

Practice with simulated trading

The various tools available to help you become more proficient are fantastic. In terms of technology, resources, and brokerage services available, it is a better time to be a trader than ever before. You can practically automate the process. The first thing to do before you commit funds to your trading activity is to do some simulated trading to get a comfort level before you actually trade.

Picking Out Your Vehicle

I'm not putting the cart before the horse, or, er, the car before the profits. Yes, you will get a chance to pick out a shiny new toy if you make wise choices in trading, but this section deals with the *trading* vehicles you put your money in. Whether you are bullish, bearish, or neutral on precious metals, there are great choices for trading vehicles. The following sections discuss the main vehicles that are very tradeable:

Stocks

Done in a stock brokerage account, stock trading can be affordable and there are plenty of precious metals and base metals stocks that are tradeable. You can probably start with a few thousand dollars and should only concentrate on a few stocks so you can watch them like a hawk. Choose them based on fundamentals and use technical analysis for your entry and exit points. Fortunately, most good brokers have a fully featured Web site and relatively low commissions. Services such as automated trailing stops and stock alerts that you can program make it easier for your trading activity. This venue can be as aggressive or as conservative as you prefer. Chapter 12 has more details for you.

Futures

Trading in futures contracts is, of course, the riskiest choice among the trade-able vehicles. For a taste of volatility and a chance to make a great profit in a relatively short time frame (in days or weeks), trading futures is the one for you. You can either be aggressive or . . . very aggressive! For beginners, a good approach would be to do spreads to limit your risk as you gain knowledge and experience. The lowest appropriate amount to do futures is risk capital in the range of $10,000 to $20,000. For more information on futures, go to Chapter 14.

Options

This is my favorite area for traders (considering all the options seminars I do, it better be!). It has both the potential for large returns and the opportunity to take advantage of volatility. Combine this with limited risk, relatively low cost, and versatility (you can do options on stocks, ETFs, or futures contracts).

Selecting Your Trading Strategy

The following sections indicate some trading strategies as they are matched up with your outlook. This is probably a better way to choose your strategy than to load you up with a batch of paragraphs (you're welcome!). I limit my examples with two tables: one on stocks with options and the second covers futures contracts with options. These tables provide a good cross-section for beginners.

Choosing your market outlook

Yogi Berra said, "When you come to a fork in the road, take it." Of course, in today's market, that fork (probably made of base metals) gives you two options: buy or sell. Everyone has an idea about "where the market is going and what to expect for that asset's next price move." There are as many trading ideas and strategies as there are traders. What is your outlook? Your strategies (for either short-term trading or long-term investing) flow from your outlook: bullish, bearish, or neutral.

Secondly, you should decide the degree to which you have your outlook. For example, say that you are bullish. That's your outlook but how bullish are you? Are you very bullish or moderately bullish? This would affect your trading approach. If you think a stock will do well then certainly, you can buy the stock. If you were very bullish on the stock then you can buy multiple call options on the stock with the same money. If you were moderately bullish, then you could buy the stock and put on a trailing stop to limit the downside while keeping the upside unlimited (of course, if the stock goes down that could trigger the trailing stop and get you out). You get the picture.

Stock trading coupled with options

Say you are looking to trade stocks (and options on stocks). Your strategies will focus on a particular stock, the Mining Trading Corp. (MTC), which is $20 per share and you have a stock brokerage account funded with $2,000. Presume that the calls and puts that you would buy (or write) are all long-term options that will expire in a year and they cost $200 per option contract. To make the math simple, presume no commissions. What could you do? Table 16-1 gives you some possibilities.

Table 16-1	Possible Choices for Stock Trading with Options	
Your outlook	**Action***	**Comments**
Most bullish strategy	Buy 10 call options on the stock (Total cost is $2,000)	If you are right, those call options would be very valuable since 10 contracts is a leveraged play on 1,000 shares of MTC. Worst case, if you are wrong, you'd lose $2,000.
Very bullish strategy #1	Buy 100 shares of MTC (Total cost $2,000)	If you are right and they go up, sell and take a profit. If they go down you lost money. Simple!
Very bullish strategy #2	Buy a call for $200 and sell a put for $200 (total outlay is $0!) The $2,000 sits as "collateral" (the synthetic long strategy)	If you are right and MTC goes up, you'd make a profit on both the call and the put. If you are wrong, then the money in your account would be used to buy MTC since writing put obligates you to buy MTC stock.
Neutral strategy with high volatility	Buy an at-the-money and put on MTC (the call straddle); total outlay is only $400	You are hoping that MTC goes either way and very far and very fast. One option would lose money but the other would be very profitable.
Very bearish strategy	Buy ten put options on the stock (total cost is $2,000)	If you are right, those puts would be very valuable since ten contracts is a leveraged bet on 1,000 shares of MTC going down. Worst case: lose $2,000.
Most bearish	Buy ten put options (total cost $2,000)	If you are correct, jackpot! If you are not, goodbye $2,000.

*Remember to use trailing stops and your broker's stock alert service where appropriate.

The above are just examples and it is not even an extensive listing. There are variations on all of the above, limited only by your education and creativity. These possibilities should whet your appetite and encourage you to look into a fascinating and potentially profitable area. Do your homework by starting with more information in Chapter 12 (stocks) and Chapter 15 (options).

Futures trading coupled with options

What kind of strategies could you implement in the futures market? Table 16-2 gives you a rundown of some strategies used by traders that are fairly simple and can be easily done in a futures brokerage account.

Table 16-2	Possible Choices for Futures Trading with Options	
Your outlook	**Action***	**Comments**
Most bullish strategy	Buy (go long) a futures contract	Absolutely bullish. Besides the risk of losing money you can also get a margin call from the broker.
Very bullish strategy	Buy (go long) a call option on that contract	Very bullish but not as bullish as item A. Since call was fully paid in cash, no margin call risk.
Moderately bullish strategy	Buy (go long) a futures contract and buy an out-of-the-money put on the same contract	You are bullish; the futures contract and the put is a hedge. If the futures contract goes down, the put would act like insurance and go up in value.
Neutral strategy	Buy a call option and a put option on the same futures contract (called the straddle)	The straddle is a neutral strategy. You are making a bullish bet (the call) and a bearish bet (put). You are hoping that it goes in either direction very strongly to make a net profit on the winning option.
Moderately bearish strategy	Sell (go short) a futures contract and buy an out-of-the-money call on the same contract	You are bearish. The futures contract and the call are hedges. If the futures contract goes up, the call would act like insurance and go up in value.
Very bearish	Buy (go long) a put option on that contract	Very bearish but not as bearish as "Most bearish" option. Because put was fully paid in cash, no margin call risk.

(continued)

Table 16-2 (continued)

Your outlook	Action*	Comments
Most bearish	Sell (go short) a futures contract	Absolutely bearish. Besides the risk of losing money you can also get a margin call from the broker.

Remember to use trailing stops and your broker's stock alert service where appropriate.

Alas, Table 16-2 barely scrapes the surface. There are many strategies and there are many combinations of futures, call, and put options. I gave you some basic strategies that are very common and very popular. Get familiar with them for starters, and use the information and resources in Chapter 14 (futures) and Chapter 15 (options) for further details.

Resources for Trading

If you take the points of this chapter seriously and apply the knowledge and information diligently, then you can make lots of money. Great! Then you can read Chapter 20 on taxes to keep more of your fortunes! But, if you aren't careful, don't worry! There is a *Bankruptcy For Dummies* guide (I might as well cover all the bases). Anyway . . .

Here are some excellent resources for traders:

- Trader Tracks (www.tradertracks.com)
- TFC Commodity Charts (www.futures.tradingcharts.com)
- Decision Point (www.decisionpoint.com)
- *Stock & Commodities Trading Magazine* (www.traders.com). This is a very comprehensive site with hundreds of products and services for traders.

For simulated trading, here are some websites that could help you:

- Marketocracy (www.marketocracy.com)
- Stock Trak (www.stocktrak.com)
- How the Market Works (www.howthemarketworks.com)

Chapter 17

Finding and Using a Broker

· ·

In This Chapter

▶ Choosing a futures broker

▶ Choosing a stockbroker

▶ Understanding and using brokerage services

· ·

*B*efore your roll up your sleeves and leap into some precious metals investments, it will be highly important to choose a broker or dealer who will help you make the transaction. Besides the fact that you don't want the broker to make you . . . uh . . . broker, a good broker is a critical part of your investing arsenal. Good brokers can save you money and make sure that your strategy is implemented. Bad brokers can lose you money and drive you crazy.

Believe me. I've been through brokers who brought tears to my eyes (and not because I was buying onion futures!). A good broker helps you make money. You have to watch out for creeps and morons. And, of course, avoid the worst of all . . . creepy morons. This chapter should guide you as you sift through the good, the bad, and the ugly when it comes to brokers.

Getting Down Some General Points

In the world of securities, you will need an individual or a firm (the broker) to help you buy, sell, and/or manage securities for your portfolio. Here is a simple breakdown for whom you need for what securities:

✔ Buying physical precious metal coins, bullion, and so on. Let me get this out of the way first. This is handled in other chapters. For bullion coins and bars, go to Chapter 10. For numismatic coins and collectibles, go to Chapter 11.

✔ If you are looking into buying mining stocks, exchange-traded funds (ETFs), or options on these securities, you will need a stock brokerage account.

✔ If you want to get into futures (commodities) or options on these speculative vehicles, you will need a futures brokerage account.

  ✔ If you are considering a managed futures account (described later in this
   chapter), that will require a futures brokerage account.

A stock brokerage account is an easy account to understand since buying
stocks is a familiar and a pervasive part of our financial landscape. However,
it can be a little confusing when it comes to an account in the world of
futures and commodities. The words "futures" and "commodities" are used
interchangeably but they are distinctly different things. Before you try to
decide on a broker, you should first decide what you will get involved with.

If you are going to speculate in the world of futures, then get familiar with the
market before you open an account. You can start with reading Chapter 14. If
stock investing is your interest, then read Chapter 12.

Both stock and futures accounts have a lot of commonality (margin, order
types, and so on) but for the sake of accuracy and orderliness, I will put them
in different segments since enough differences exist. The first thing to tackle
is a futures brokerage account.

Futures Brokers and Accounts

If you are going to speculate with futures and options on futures, you will
need a futures brokerage account. You will read a lot about investing in the
literature and Web sites and I am even using that term in context, but please
notice that I try to use the term "speculate" and not "invest" since this is an
area that is high-flying, volatile, risky, and can in some ways like financial
gambling. Even if you are very comfortable and proficient with futures and
options on futures, you should devote only a relatively small amount of your
total financial assets in this roller-coaster world.

I find it okay to price-shop brokerage accounts for stock investing but I don't
feel the same way about futures. It's important to have a professional broker
who is experienced with the ebb and flow of this market. You want someone
who has integrity and who can implement your trades according to your
strategy. In terms of how they relate to you, there are two types of futures
brokerage accounts, full-service and discount.

Full-service futures broker

In a nutshell, you will pay more for a full-service broker but you will get more
services. In futures, a broker with experience, expertise, and resources can
be valuable. The broker involved works with you to help you choose strategy
and implement transactions. He or she offers advice and research on the
various markets. The full-service futures broker also provides feedback on
developments that can affect futures in general and the positions in your
account in particular, such as the day's news and events.

Discount futures broker

The discount brokerage firm will basically place your orders for you with no guidance or advice. You get fewer services so that you end up paying less in fees and commissions. Ultimately you will do your own research and due diligence. A discount broker is a good choice for confident, experienced individuals who just want an inexpensive way to trade.

Both types of brokers usually offer good Web sites with content and links that can help you be more informed with your speculating. As a financial planner who does all of his own, hands-on investing, I am very comfortable with a stock discount broker but I prefer a full-service futures broker. I believe that beginners and intermediate level investors and speculators are better off with a full-service broker because of all the reasons and risks mentioned in this chapter as well as in Chapter 14. And hey . . . if you get really good at it you can always move on to discount brokers.

Let me put another wrinkle in this topic about the different types of brokers. Whether the broker is full service or discount, you should know something about what type of brokerage firm it is from an industry and regulatory perspective.

The futures commission merchant (FCM)

The FCM is an organization that provides the principal order clearing services for the futures industry. In other words, they are the main organizations through which the buy and sell orders of futures contracts (and options on them) are made. FCMs are members of the individual futures exchanges such as the New York Mercantile Exchange (NYMEX) and the Chicago Board of Trade (CBOT), among others. You can actually see a complete listing of FCMs at the Web site of the Commodity Futures Trading Commission (www. cftc.gov).

Introducing broker (IB)

Unless you are working with the retail division at the FCM, you are likely to work with an introducing broker (IB). The IB can be a full service or a discount brokerage firm that works through the FCM. The transactions done by the IB are cleared through the FCM. The IB may or may not be a recognizable name but it provides the service you seek (either full service or discount). Although IBs can be large firms, they tend to be smaller firms which means the service can be more personalized.

Selecting a Broker

Unless you get a referral to a broker from someone you trust, such as a close friend or your tax or legal advisor, you'll end up doing some research to find a broker for your futures business. There are several places to start.

Commodities Futures Trading Commission (CFTC)

The CFTC has consumer information on futures and futures brokers. You can visit them at www.cftc.gov.

National Futures Association (NFA)

I highly recommend that investors visit their Web site at www.nfa.futures. org for several reasons. They do provide information and guidance on choosing brokers. In addition, you can find out if there are any complaints or problems with brokers in question. If you have been victimized, you can file a complaint. Even though it is a private entity, the NFA does work as a regulatory body that can discipline wayward brokers.

In addition, the NFA does have educational resources and links for the investing public.

Futures Industry Association (FIA)

The Futures Industry Association (FIA) is the trade group for futures brokers. It offers a wealth of information at its Web site (www.futuresindustry.org), including an extensive directory of brokers, futures information services, and links to futures-related companies, programs, and educational resources.

The exchanges

The futures exchanges also provide information on choosing a broker. A good example is the Chicago Mercantile Exchange. It runs a service called the "Find a Broker Program" at its Web site, www.cme.org. Other exchanges that have lots of useful information are

- ✔ Chicago Board of Trade (www.cbot.com)
- ✔ New York Mercantile Exchange (www.nymex.com)

Keeping your eyes peeled for ICE

When you are selecting a broker, you are basically choosing someone for your financial team in the same way you choose a tax advisor or insurance broker. It will hopefully be a long-term relationship. You obviously need someone whom you are comfortable with. Let me share with you what I look for when I'm meeting people whom I hope to do business with. I look for people who have ICE in their veins. What is ICE?

Integrity: They must have integrity or why bother working with them?

Competence: If they are true professionals, then they should know what they are doing.

Enthusiasm: Do they enjoy what they are doing? If they don't, how will they get good at it? If they are not enthusiastic about their work, there's a good chance they won't be doing this line of work in due course.

After going through several brokers and firms, I finally found my broker Charlie. He's been my futures broker since 2003. He treated me well and it paid off for him because I sent him dozens of clients. In my book, he has ICE in his veins (kinda catchy, isn't it?).

That's probably a good lesson for anyone in business. If you treat people well, they will not only stick with you but they will also send you business.

Dealing with Futures

I personally like options on futures versus futures directly. Yes . . . find out as much as possible about futures, but options on futures offer much of the same potential but with less risk. When I buy an option on a futures contract and the futures contract goes down the next day, I don't worry about a margin call. And if it's a cloudy day . . . I stay home and play Parcheesi. The sections that follow should give you a good look at choosing and using a futures broker.

Interviewing a futures broker

After you get past "What's your name?" and "How 'bout those Mets!" you start getting into that uncomfortable silence. Have no fear! This Dummies guide is here. Here are some questions that should merit some answers:

✔ How many years of experience do you have working futures? You would like three or more years (but not 87).

✔ Do you have an area of futures that you specialize in? Some brokers specialize in metals; for others it is currencies or grains. Make sure they have proficient knowledge in your desired markets.

✔ Does your approach embrace fundamental analysis, technical analysis, or a mix of both? You should know something about their knowledge of these areas and what they favor as an approach.

✔ What size account are you comfortable working with?

✔ What are your firm's requirements for margin? Can you give me some examples of how margin would work with my account?

✔ What are the commissions and fees and what services are provided?

✔ What is your track record (provided they have done discretionary trading for clients)?

✔ How are disputes or concerns handled at this firm?

Margin in a futures account

Margin in a futures account is a tad more complicated than margin in a stock brokerage account. Well . . . a big tad indeed.

For the futures broker, margin is a performance bond. In the area of futures, the volatility in market prices from day to day can be great. Since you put down money roughly equal to 20% (sometimes more) of the futures contract, it is possible that the futures contract market price goes against you. The margin in a futures account acts as a good faith deposit to cover swings in the market value of the contract. If the market price change is significant and adverse, you may be required to put more money in the account (the margin call) to maintain the necessary margin. Keep in mind that in futures, the margin you put down is not a down payment on a futures contract. A futures contract is not an asset you own; it is a liability, especially if the market moves against you. The margin is a good faith amount sitting in your account just in case you are wrong about the performance of the futures contract market price.

The margin is effectively an estimate of your possible loss in the next trading session (the next day) escrowed in advance. Because no one knows in advance which way the market could move, margin is posted by both the buyer and the seller.

Because the rules for margin can differ from broker to broker, work through some what-if scenarios with your chosen broker given the market you are looking to trade.

Avoiding problems in your account

Churning is a problem that you need to watch out for. *Churning* is when a broker performs a high level of activity (buying and selling) in the account for the (unspoken) purpose of generating commission income. Sometimes it doesn't seem evident in a commodities account since it can be quite normal to see lots of trading transactions in a relatively short period of time. You must be most wary of churning especially when you have given the broker discretionary trading authority in that account.

A new client came to me after closing out an account with a dubious firm located in Florida. He started the account with $60,000 and got out four months later with $57,000. Getting out with a net loss of $3,000 is not the end of the world . . . but that's not the story. The real eye-opener came when Charlie (my broker) analyzed the account's activity. The statements indicated over $17,000 of commissions during that brief four-month period! Gee . . . maybe if my new client was so concerned with making a killing he should have gotten a job there instead.

In addition to churning, the most often cited complaints have been high commissions and poor service. There are many brokers and you can shop around. Once you find a broker whom you are comfortable with, then it's time for the fun stuff: filling out forms.

Opening a futures account

Here are the steps in finally opening your futures account:

1. **Read the forms and ask the questions (see the section above).**
2. **Put in the usual information: name, address, and so on.**
3. **Decide on the initial investment.**
4. **Understand what risks you will tolerate.**
5. **Read the disclosure agreements about risk.**
6. **Submit the application and your funds and wait for the green light to start trading.**
7. **Once you are ready, remember to read (re-read?) Chapter 14 on futures.**

After you have done your due diligence, you can now open up a futures account. When you see the paperwork and supporting documentation, you'll think that they cut down an entire redwood tree just for the forms (makes me regret not going long on those lumber futures!). Anyway, after the usual stuff about your name, address, social security number, and so forth, you'll see questions and forms covering the topic of risk.

Commodities regulations try to make sure that small investors are made fully aware of the risks associated with futures and options on futures. You'll see for our purposes that there are really two different categories of investors in the world of futures (again, it's really speculating): the retail investor and the accredited investor.

The retail investor is the little guy or the small speculator (technically called the non-accredited investor). If you have a regular job or you run a small business and your income is less than $200,000, then you are a retail investor. This category covers most folks. This is the category that the authorities (such as the Commodity Futures Trading Commission) attempt to give some protection through documents that need to be filled out when you open your futures account. You'll be required to sign risk disclosure documents so that you are made aware of potential risks.

The accredited investor is a different animal. It may not necessarily mean that you are particularly proficient at futures but it does indicate that you can sustain losses without undue impact on your financial situation. An accredited investor is generally defined as one with total financial assets of $1 million or more (not including the value of your residence) and/or income exceeding $200,000 per year.

Futures account commissions and fees

The costs of transacting futures is a major issue. Asking your potential broker about how much he or she charges in commissions is necessary and obvious. Choosing on price alone is not always the wisest thing since you want to make sure that you are getting value. Let me give you an example.

My futures broker is Charlie and I think he's great. His commissions are not the cheapest and they're not the most expensive; the commission price is in the middle of the pack. I could certainly find cheaper commissions elsewhere. Heck . . . just go on the Internet and you can see bare-bones commissions. However, we can't confuse price with cost. I may pay $50 for a trade and someone else may only pay $10 because it's an Internet trade. Yet, more times than not, I will save money over that Internet trade. Why? In my case, Charlie "shops the price" as he places the order with the folks on the exchange floor. He will end up buying the trade at a good price and save me $100 or more on the total trade. This is not easy to do with a cookie-cutter approach at a futures Web site since the Internet approach may lock in a price without negotiating.

Here are more points about commissions and fees:

- ✔ **Round-turn commissions:** A round-turn commission covers both the buy and sell sides of the trade. This is typical when you purchase an option on futures.

✔ **Half-turn commissions:** A half-turn commission only covers the buy or sell side of the trade. This is typical when you purchase futures directly (versus options on futures).

✔ **Negotiable commissions:** Push for lower rates, especially if you are (or will be) an active trader.

As of early 2007, average commissions per half-turn are in the range of $30 to $80. If you are in the upper part of the range, negotiate.

If you are using a full-service broker whose guidance has lost you money, consider negotiating for lower rates or (if they really do a bad job) then move your business elsewhere.

If you are using a full-service broker and they are helping you make money then don't be that concerned about the commission. Good guidance is hard to find so why ruin a good thing?

All in all, commission rates are a better deal now in the Internet age versus what they were 10 or 20 years ago. During the 1990s, many futures brokers charged $120 to $150 or more.

Remember that whatever person actually helps you in the trade, that person will not be getting 100% of the commission amount. It does get divvied up by several parties. If that person is an introducing broker then he or she must pay roughly $10 to $15 from that commission to the organization clearing the trade, the FCM.

Futures orders

In the world of futures, there are many different orders you can implement and I'll try to list as many as space can accommodate, but the most common trade orders that individuals will come across are market orders, limit orders, and stop orders, so they'll be covered first.

In speaking to your broker to put in an order, he/she will take your instructions and repeat the order back to you before it is submitted. Remember that even though it is a verbal order, it is effectively a contractual transaction. The odds are that the conversation will be recorded just in case there is a question or a dispute regarding the verbal order.

Once the order is confirmed between the client and the broker, the broker will then relay the order by telephone (or electronically) to the broker's representative on the trading floor at the futures exchange (more on exchanges in Chapter 14).

Before we get to the transactional orders (like market orders), let's first mention time condition orders.

Time condition orders

This is a simple order and it goes in conjunction with the transactional orders. The time order is your order to make the transactional order either a day order or a good-'til-cancelled or GTC order. In other words, you will tell the broker that your order is in effect either for the day or if it will be in effect indefinitely.

In a day order, you may say, "Buy a December silver futures contract at the price of $X or better and this order is good for the day." This order will be filled only if silver hits $X (or a better price) during that day's trading session. If it doesn't hit $X by the end of the trading day, then the order will not be executed and it will expire and be cleared off the slate. That's a day order.

A GTC in that same situation may be, "Buy a December silver futures contract at the price of $X or better and this order is good 'til cancelled (GTC)." With a GTC order, if that December silver futures contract doesn't hit the price target that day then the order will stay on during the days to come until it gets filled. The GTC order won't stay there forever. Either it is filled, or the client cancels it, or it may expire based on a time frame designated by the broker. Depending on the broker, the GTC order can remain active (open order) for 30, 60, 90 days, or other time frame (check with your broker).

That wasn't so bad. Remember that if you are not sure if you are placing the order properly, don't be shy; tell the broker what you are trying to accomplish and ask for assistance in understanding and implementing the order. Hey . . . it's your money and you're paying for service.

Market orders

This is the most common transactional order. In a market order, the broker tries to fill your order as soon as you submit it. A market order could be voiced by the client as, "For my account #1234 buy 2 Gold April futures contracts at the market."

The market order is a simple and common order and I use it frequently. Understand that a market order won't always get you the best price because in the order you are basically saying that you will accept the market's price instead of waiting for a better price. For me, if I'm getting into a transaction worth thousands of dollars, I won't sweat $50 either way because if I dig in my heels to wait for a $50 savings, I may miss out on the order altogether.

Limit orders

In a limit order, the client requests that the trade be done at a specific price (or better). An order may be, "Buy 1 platinum July futures contract at $75 or better good for the day." It may be filled at $75 or you might get it at a better price like $74.50 or $74. But it won't be filled at a price above your specified price. Limit orders are good to use during weak or quiet markets since it will be easier to obtain your price. A limit order can be a day order or a GTC order.

Stop orders: Buy and sell stops

A stop order is an order that turns into a market order the moment the market price of that futures contract hits a specified price. More accurately, stop orders can be buy stop orders or sell stop orders.

The buy stop is placed above the market price and it becomes a market order when the futures contract trades at or above the specified stop price.

The sell stop is placed below the market price and it becomes a market order when the futures contract trades at or below the specified stop price.

Cancellation orders (CXL orders)

Also called a straight cancel order, this order instructs the broker to cancel an order previously entered. That was simple!

Cancel former order (CFO order)

This order basically does the work of two orders. It cancels a prior open order and then replaces it with a new order.

Market on close order (MOC order)

This order instructs the broker to fill the order during the closing of market trading (during the last 30 seconds). The order must be filled at a price that is within the closing range. Whew! I wouldn't want to be a broker missing this order by seven seconds with a temperamental client.

Opening-only orders

This order gets implemented during the opening of trading that day. As with the MOC order, the order is transacted within a range of prices as the market opens.

One cancels the other order (OCO order)

This is technically two orders entered simultaneously but there is a condition. The condition is that if one order is filled then it automatically cancels the other order. Order #1 may be the desired order to be filled but if market conditions make it impossible to fill, then order #2 would be filled. The filling of either order means that the other order is not filled. Say that fast three times. Made ya try! Well, this order is akin to "If it's sunny today, we're going to the monster truck rally and get rowdy but if it's cloudy we'll stay home and play Parcheesi."

Other orders

There are other orders as well and for the sake of completeness, I'll just list them here. Speak to your broker about their applicability to you and your account:

- Market-if-touched orders (MIT order)
- Fill-or-kill orders (FOK orders)
- Disregard tape orders (DRT orders)
- Wire orders

Most brokers list the various orders that they can transact with complete description (or a glossary) and examples in their literature or at their Web site.

Managed futures accounts

Managed futures accounts have been around since the late 1970s. A managed futures account is a futures account that is managed by a professional money manager registered with the Commodity Futures Trading Commission (CFTC) and officially called a commodity trading advisor (CTA).

The CTA is a regulated professional who meets the educational requirements of the National Futures Association (NFA), a self-regulatory industry watchdog organization. (It operates similarly to the National Association of Securities Dealers [NASD].) Most CTAs use a proprietary trading system or some other formal method through which they make their trading decisions. CTAs could go long or short futures contracts.

A managed futures account has the following advantages:

- Diversification: For investors with money in stocks and bonds, commodities add a new dimension to their portfolio.
- Professional management: You don't have to choose the individual futures to include; the CTA makes the day-to-day decisions.
- Protection against political crises and natural disasters.
- A hedge against inflation (such as rising food and energy costs).

An important aspect of commodities is that they do perform quite differently from stocks and bonds so they add value to your overall portfolio. Many studies over the past quarter century have shown that commodities as a general investing class have performed very well during periods of political and economic stress.

Commodities performed very well during the economic and political turmoil of the late 1970s and again during the first half of this decade in the wake of 9/11, hurricanes, and rising inflation. Diversified managed accounts were among the best performing accounts in those time frames.

If you are in (or considering getting into) a managed account, ask the CTA about the approaches used to maximize profit and minimize risk. Here are the most common approaches used by CTAs:

✔ **Going long:** Making a bet that something will go up in price.

✔ **Going short:** Making a bet that something will go down in price.

✔ **Spreads:** A hedging approach where both a long and short position are held. More about spreads in Chapter 15.

Some considerations about CTAs

You don't have to blindly accept anything the CTA says. Their track record and methodology can be reviewed.

Drawdowns

This is important information regarding the CTA's track record in the event that there was a decline in the equity (or account value) of a managed account. A drawdown represents the maximum downward move of the account value from the peak to the trough or valley. It is like a worst case scenario taken from the CTA's recent activity. It is not meant to make you think that the same dip in account value is going to happen to you but it is meant to indicate that losses can happen. It can also indicate how long losses can be recouped as the market rebounds. Past performance is included in the disclosure documents as required by the National Futures Association (NFA).

The disclosure documents also indicate how the performance did over an extended period of time and also show annualized percentage gains and losses to make it easy for prospective clients to compare with other CTAs and with acceptable market averages.

Dispersion

Dispersion is the distance of CTAs' monthly and annual performance using a mean or average level, which is a common way to evaluate how well they have done.

Fees and account minimums

A managed account by a CTA is not like a mutual fund or hedge fund. The client has the ability to review the account and see what trades are made. CTAs don't make money from commissions since that would be a conflict of interest. The account's trades are cleared through a FCM (see above). CTAs typically make money as a percentage of the account's performance. Minimum account sizes can vary but they have been as low as $25,000 although most CTAs require $50,000 or more.

Managed futures accounts are not for everyone. Small investors are better off looking elsewhere (such as the stock market). But for those with large portfolios who want commodities in the mix, managed accounts are a viable choice.

They must be doing something right; managed futures have over $150 billion under their guidance. As inflation heats up and more of the world's population needs more of the basics of life, commodities will continue to be an attractive alternative. To find out more about CTAs, head over to the NFA's Web site at www.nfa.futures.org.

Stock Brokerage Accounts

After slogging through all that stuff about futures accounts, this should be a breeze. Many folks have opened a stock brokerage account and it's probably the most appropriate for most small and mid-size investors (or speculators).

For most investors seeking growth (either conservative or aggressive) without the volatility and risk of futures, stocks and ETFs (and options on these securities) offer plenty of variety. A stock brokerage account should actually be the cornerstone of the long-term investor's wealth-building program because of the wide array of investments, techniques, and strategies available. There are several types of brokers . . .

Stock brokers

A long time ago, there was one category of stockbroker. That category was called . . . stockbroker! Then things changed during the 1980s. For investors like you and me, there came to be two basic categories:

- ✔ **The full-service broker:** For investors who wanted more attention and service from their broker, a full-service broker was the choice. The commission was high but many investors felt the price was justified due to personalized service and more guidance in the realm of choosing investments and setting financial objectives. If you think you need more attention and assistance and don't mind the extra cost, the full-service broker is for you.

- ✔ **The discount broker:** With deregulation in the early 1980s, this paved the way for greater competition in the stock brokerage industry. With trailblazers like Charles Schwab, the discount broker was born and soon became a new and growing force in the world of stocks (and futures).

Account types

As you get into stock investing, you'll find that there are three basic types of accounts: cash, margin (or margin and options), and discretionary. Retirement accounts (such as Individual Retirement Accounts, or IRAs) are covered in Chapter 20.

Cash account

This is the simplest and easiest account to open. In a cash account you pay everything in cash with no credit involved. This one is appropriate for small investors or for conservative investors. As you shop around, you will find stockbrokers who take accounts with initial minimum investments of $2,000 or less.

Margin account

A margin account basically means that you have borrowing privileges in your account. In other words, you have the ability to use securities in your account as collateral and use the loan proceeds to purchase other securities in your account.

Other accounts

Another type of account is a discretionary account where you give trading privileges to others to trade your account. Of course you have to do some serious due diligence before you do this, but I think that you are better off doing your own research through the resources mentioned throughout this book.

Lastly, just about every stock brokerage has the capacity to help you open Individual Retirement Accounts (traditional or Roth). They are covered in Chapter 20.

Opening a stock brokerage account

It's not difficult to open up a stock brokerage account but just to be sure, here's a step-by-step rundown:

1. **Do the research.** There are plenty of sources that evaluate brokers. You can head down to the library and look up *Forbes* magazine and *Barron's* that do regular reviews and comparisons of brokers. Web sites such as www.smartmoney.com, www.investopedia.com, and www.sec.gov also give great consumer information on choosing a broker.

2. **Ask your questions.** What are the commissions? What is the minimum investment? The sources in Step #1 can help with your inquiry.

3. **Choose the type of account.** A cash account is easy to get but you're better off getting the margin and options account open so that you have more choices (such as buying/selling calls and puts).

4. **Get the paperwork.** Either call the broker or visit their Web site. The forms can be downloaded and mailed (after you fill them out, of course). Many brokers let you fax the application or fill them out online at the Web site.

5. **Fund it.** Either mail in a check for the initial investment amount or transfer money electronically from your bank account.

6. **Make a truckload of money!** Well . . . let's hope so.

Once you've got all that, you're ready for business! Keep in mind that just about all the broker Web sites have complete details on how to do an order and so on. Of course, you can call customer service at their ubiquitous 800 number.

Types of Orders

When you are getting into stocks, futures, or options, how you invest can be just as important as what you invest in. Of course, the "what" is precious metals (and securities related to them) but "how" you do it is important as well. When you make that buy or sell transaction, you will need to be familiar with what type of order. Doing the order through the broker's Web site is your best bet.

- ✔ **Market order:** This one works the same as with the futures broker. A market order is the instruction to the broker to purchase (or sell) that security immediately.

- ✔ **Buy limit order:** The order to purchase a security at a specified price (or better).

- ✔ **Sell limit order:** The order to sell a security at a specified price (or better).

- ✔ **Stop-loss order:** An order to sell a security you purchased previously at a specified price below the market price only in the event the security is dropping in price. A stop-loss order has the purpose of minimizing the downside risk.

Stock brokerage services

In today's market and with today's technology, brokerage service is no longer merely a way to buy and sell securities. The stockbroker (even a discount one) has features and bells and whistles that can make for a more profitable investing experience. In the prior segment, I mention stop-loss orders. Now take it a step further with the trailing stop strategy.

Trailing stops

For those of you who read *Stock Investing For Dummies* (pat yourselves on the back) you will know that I am a big proponent of trailing stops. I spent a lot of time on the advantages to investors that brokerage services can provide (such as the trailing stop). I don't use them in every circumstance but I consider them an important risk-management tool in the investors' arsenal.

What is the trailing stop? I'm glad you asked! It is an active strategy of implementing stop-loss orders. The stop-loss order (see above) is an order to help you reduce or remove the downside risk to holding a stock. A trailing stop means that you keep adjusting the stop-loss order to protect more and more of your investment as it increases in value. Let me provide an example.

Presume that you have 100 shares of Eureka Gold Mining Company (EGM) at $30 a share. Let's say you put on a 10% stop-loss order and you make it GTC (good 'til cancelled). In that case, the stop loss would be at $27 and would stay on indefinitely (based on how long that particular broker decides a GTC order time frame would be). Remember that if EGM's stock price falls, the worst that would happen is that it could hit $27 and be sold immediately and you wouldn't have to see EGM go to $26 or lower. What happens when EGM does what you hope it does and goes up?

Say EGM goes to $40. This is great. So what happens with the stop-loss order? At this point you should cancel that stop loss at $27 and replace it with one at, say, $36. That would make it again 10% (10% of $40 is $4; $40 less $4 is $36) and again it would be a GTC order. At this point, you are protecting the original $30 per share plus a profit of $6. Again, the upside would have no limitation while we are limiting the downside. To protect 100% of your original investment plus a nice chunk of profit is a good deal.

Broker e-mail alert services

Do you know that you could program your account at the broker's Web site as if it were a computer?! Sounds almost futuristic but it's really cool. Here's a brief rundown of what you can do:

- ✔ **Price alerts:** If a stock you have (or one you're interested in) hits a price you specify, an e-mail will be sent by the broker to let you know.

- ✔ **News alerts:** When something happens such as major market-moving news, an e-mail will alert you.

- ✔ **Trade pre-programming:** You can go to the Web site and put in orders to buy or sell a stock, ETF, or option when certain conditions are met (when that security hits a price or when another event occurs). An e-mail will then be sent to let you know it happened.

- ✔ **Trade triggers:** Being able to set up orders to occur when primary or secondary events occur. An example would be "buy or sell security B when security A does something (such as reach a 52-week high)."

The things you can do are limited only by your imagination. The point is that these added services give you more control over your securities in today's marketplace. These are just more tools in your wealth-building arsenal.

If your broker doesn't yet have these services, it's okay. There are some great Web sites that can give many of these features. A good example is Market Watch (www.marketwarch.com).

Stock brokerage commissions and fees

The obvious charge for your trading comes in the form of commissions. But fortunately, with the advent of the Internet, most of the discount brokers give you the ability to buy or sell securities ranging from $7 to $15, making stock investing affordable for almost anyone.

Margin in a stock brokerage account

Using margin in your account can increase the positive . . . or increase the negative. After you have been approved for margin trading, you will be allowed to borrow against marginable securities. A marginable security is basically almost any stock that is not an over-the-counter stock or a small stock on Nasdaq. Most stocks listed on the New York Stock Exchange, for example, are marginable.

With margin, you could, say, put $2,000 in an account and be able to buy up to $4,000 in stock. The minimum amount to set up a margin account will depend on the broker (it could be for $5,000 or more). When you initially put that $2,000 in the account, you will see (at the broker's Web site, for example) that you have cash of $2,000 and buying power of $4,000. (The broker lends you the difference.) With listed or marginable stock, the borrowing ratio is 50%. The ratio could be different depending on the security. (You can borrow up to 90% of the market value of Treasury bonds, for example.) Keep in mind that you will pay margin interest on anything you borrow through your account.

The use of margin can magnify gains or . . . if you're not careful . . . losses. Say you decide to start with $3,000 and the stock you're going to buy is 100 shares of the Hokey Smoke Gold Company (HSG) at $60 a share. In this case, you would use the full limit of 50% to make the purchase. The 50% is the limit and you don't have to go that high. It could certainly be less. Say that HSG goes up to $75 per share making the total stock value $7,500. How did you do? That $15 gain amounted to a 25% gain ($15 is 25% of $60, the original stock price) but since you used only $3,000 of your own money your actual percentage gain is 50%. Cool! At that point you could choose to sell the stock and pay off the margin (don't forget margin interest and commissions). It's nice if it happens . . . but . . . what if you're wrong?

Let's say that HSG's stock price drops to $40 (Hokey smoke!!). Now what? The account is now worth $4,000 but $3,000 is the margin loan from the broker; the loan ratio has fallen to 25% (your equity portion of the $4,000 account value is only $1,000 or 25% of the total amount). This is a problem and you will get a margin call to get that ratio back to 50% within one trading day. You would either have to kick in another $2,000 or deposit more marginable securities in the account to restore the ratio to 50% (or better). If you don't then the broker can liquidate positions in the account to restore the ratio. This is in keeping with the agreement you have with the broker which says that you must maintain your margin ratio at the required level (or better).

To stay out of trouble with margin, here are some reminders . . .

- ✔ With margin, you will be charged interest. The longer the borrowing period, the more you will pay.

- ✔ Margin can magnify losses as well as gains.

- ✔ Some stocks are eligible for margin trading and some are not. Check with your broker for more details.

- ✔ The money that you receive from selling stocks that were initially bought on margin will first go into liquidating any margin loans outstanding before you have any funds to withdraw or reinvest.

- ✔ Different brokers have different requirements regarding margin so, again, check with the broker about terms and conditions.

- ✔ Novice investors are best advised to use little or no margin to avoid the risks and volatility that can occur. Wait until you accumulate more funds and . . . more importantly . . . more experience and expertise.

Margin interest can be tax-deductible as investment interest so discuss it with your tax advisor (more on taxes in Chapter 20).

Chapter 18

Using Technical Analysis

*W*hen figuring out what to do in the investment world, most professionals use one of two basic approaches: fundamental analysis or technical analysis (many use some combination of the two). Both approaches are used in a number of markets ranging from the stock market to commodities. I'll limit the comments to our narrow segment of the financial world, precious metals in the forms of futures (and options on futures). Of course, fundamental and technical analyses can be applied to individual mining stocks, exchange-traded funds, (ETFs), and on precious metals indexes (such as through options). Generally, technical analysis is not used with some precious metals vehicles such as bullion.

In regards to precious metals futures, fundamental analysis goes into the economics of the underlying investment such as supply and demand information as well as factors affecting the investment such as politics and regulations. (If the investment in question was a company then you would include analysis on its financial strengths and profitability.) Technical analysis tries to understand where the investment's price is going based on market behavior as evidenced in its market statistics (presented in charts, price, and trading volume data). This will all be fleshed out in this chapter of course. Technical analysis doesn't try to figure out the worth of an investment; it is used to figure out where the price of that asset or investment is trending.

Because technical analysis is indeed technical, don't try to understand it all immediately. What normal person gets into stochastic oscillators and divergent and convergent who-z-what-is the first day?! "Not I!" said the author ("Amen, brother!"). Understanding technical analysis is like eating an elephant: You eat one bite at a time (and you put the rest in the trunk for later).

Technical analysis is most useful for those who are trading and/or speculating during a relatively short time frame measured in days, weeks, or months. It is not that useful when you are trying to forecast where the price will be a year or more down the road.

Technical versus Fundamental Analysis

Although technical analysis is "the star" of this chapter, it is useful to take in its shortcomings and juxtapose it with fundamental analysis. The first and major drawback of technical analysis is that it is a human approach tracking human behavior in that particular market. In other words, just because it is called technical analysis doesn't mean that it's technical a la the law of physics. It is called technical analysis because the data you look at is technical but the movement of the price of the underlying asset or investment moves due to the cumulative decisions of many buyers and sellers who are human and therefore fallible. Why mention this?

Everyone is looking to make money and there are many trading systems and approaches that are based on technical analysis. Unfortunately, making profitable investments isn't just 2+2 = 4. If technical analysis made things so easy that mere computer models or trading systems could give you a voila-money-making decision then everyone could and would do it. Yet, that is not the case. Let me give you my take.

I favor fundamental analysis for long-term investing. I have shunned technical analysis for choosing individual stocks because I didn't see the long-term value in it. In stock market history, if you did a nose count of successful investors and what approach they used, you would find that those long-term investors who used some variation of fundamental analysis (as those who use a value-investing approach) are overwhelmingly the larger category. The legendary investors such as Warren Buffet and Peter Lynch rarely looked at a chart. But before you throw out the technical analysis with the bath water, read on.

Those who use technical analysis in short-term trading or speculating in larger scope investments tend to do better than those who don't use it. What does that mean? What I mean is that if you apply technical analysis in something larger than a company such as an index or a commodity, then you will tend to do better. If you are getting into trading futures on entities such as grains, or energy or precious metals, then understanding the basics of technical analysis will make you, overall, a better (hence more profitable) trader. Since short-term market behavior and psychology can be very mercurial and irrational (human), then technical analysis will have usefulness.

The guts of technical analysis

When you are using technical analysis, understand how it operates and what it is that you look at. Technical analysis, for purposes in this book, is based on several assumptions:

The price tells all

The asset's market price provides enough information to render a trading decision. Those who criticize technical analysis point out that it considers the price and its movement without adequate attention to the fundamental factors of the company. The argument made favoring technical analysis is that the price is a snapshot that does reflect in and of itself the basic factors affecting the company; that includes the company's (or asset's) fundamentals.

Technical analysts (also called technicians) believe that the company's fundamentals, along with broader economic factors and its market psychology, are all priced into the stock, removing the need to actually consider these factors separately. The bottom line is that technicians look at the price and its movement and extract from this a forecast for where it is going.

It's all about the trend

The price tends to move in trends. In the world of technical analysis, the phrase "the trend is your friend" is as ubiquitous as the phrase "you spoiled the broth, now you lie in it!" is in the restaurant industry. Uh . . . maybe more so. Following the trend is a bedrock principle in technical analysis and the data is either there to support the trend or not. When a trend in the asset's price is established, the tendency is that it will continue. The three types of trends are up, down, and sideways (but you knew that).

History repeats

Another foundational idea in technical analysis is that history tends to repeat itself, mainly in terms of price movement. The repetitive nature of price movements is attributed to market psychology; in other words, market participants tend to provide a consistent reaction to similar market stimuli over time.

Technical analysis uses chart patterns to analyze market movements and understand trends. Although many of these charts have been used for more than 100 years, they are still believed to be relevant because they illustrate patterns in price movements that often repeat themselves.

How about both?

I think that a useful way to combine both is to use the strength of each. Fundamental analysis should help you understand "what" to invest (or trade or speculate) in while technical analysis guides you as to "when" to do it. Since markets ebb and flow, and they zig and zag, technical analysis can help you spot low-risk points to either enter or exit a trade. Technical analysis, therefore, helps to stack the deck a little more in your favor. Considering how markets are going lately, every little bit helps.

Blending the two approaches to some extent has been done with success. Obviously, if the fundamental and the technical factors support your decision then the chance for a profitable trade has more going for it. How could this blend occur?

For example, look at the concepts of oversold and overbought (see the segment on Relative Strength Indicator later in this chapter). If you were looking at buying a stock (or other asset) because you thought that it would be a strong investment but were not sure about when to buy, you would want to look at the technical data. If the data told you that it was oversold then it would be a good time to buy. *Oversold* just means that the market was a little too extreme in selling that particular investment during a particular time.

By the way, I like to think that the technical terms oversold and overbought have a parallel to fundamental terms such as undervalued and overvalued. Since fundamental analysis is a major part of a school of thought referred to as value investing, the concepts make sense (yes . . . I am into value investing). Just as it is usually a good idea to invest in an undervalued stock so it is a good trading idea to buy a stock that is oversold. It is logical to presume that an oversold stock is undervalued (all things being equal). Of course, the other terms (overbought and overvalued) can also run in tandem. I might as well finish here before you are overwhelmed and under-interested.

On the other hand, the fundamentals can help a technical analyst make a better trading decision. Say that technical analyst has a profitable position in a mining stock called Six Feet Under Co. (SFU). If the technical indicators are turning bearish and the new quarterly earnings report for SFU indicates that net profit is significantly lower, then selling SFU's stock is probably a good idea.

The tools of the trade

When you roll up your sleeves and get into technical analysis, what will you be dealing with? It will depend also on what type of technical analyst you are. In technical analysis, there are two subcategories: those who predominantly use charts (these technicians are called . . . chartists!) and those who predominantly use data (such as price and volume data). Of course, many technicians use a combination of both.

> **Charts:** The neat pictures that graph price movements (such as chart patterns)

> **Data:** Such as price and volume information (along with technical and behavioral indicators derived from it)

Technical analysts don't look at the fundamentals because they believe that the marketplace (as depicted in the charts, price, and volume data) already takes into account the fundamentals.

Tracking the Trend

Identifying the trend is a crucial part of technical analysis. A trend is just the overall direction of that security or commodity. Which way is the price headed? It's easy to see which way the asset is headed in Figure 18-1.

Unless you are a skier, that's not a pretty picture. The bearish trend is obvious. But what do you do with a chart like Figure 18-2?

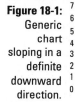

Figure 18-1: Generic chart sloping in a definite downward direction.

Figure 18-2: Generic chart showing a sideways pattern.

Yup . . . looks like somebody's heart monitor while he or she is watching a horror movie. A sideways or horizontal trend just shows a consolidation pattern that means that the security or asset will break out into an up or down trend. Regardless of whether the trend is up, down, or sideways, you will notice that it is rarely (closer to never) in a straight line. The line is usually jagged and bumpy since it is really a summary of all the buyers and sellers making their trades. Some days, the buyers have more impact and some days it is the sellers' turn.

Keep in mind that there are formal definitions of the three basic trends.

✔ **An uptrend or bullish trend** is when each successive high is higher than the previous high and each successive low is higher than the previous low.

  ✔ **A down-trend or bearish trend** is when each successive high is lower that the previous high and each successive low is lower than the previous low.

  ✔ **The sideways trend or horizontal trend** shows that the highs and the lows are both in a generally sideways pattern with no clear indication of trending up or down (at least not yet).

Figure 18-3 shows all three trends.

Technical analysts call the highs peaks and the lows troughs. In other words, if the peaks and troughs keep going up, that's bullish. If the peaks and troughs keep going down, it's bearish. And if the peaks and troughs are horizontal then you are probably in California (just kidding).

Figure 18-3:
Chart that simultaneously shows an up, down, and sideways trend.

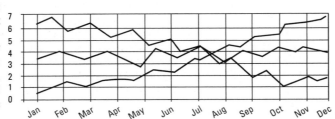

Trend lengths

With trends, you are not just looking at the direction, you are also looking at the trend's duration or length of time that it is going along. Trend durations can be (you guessed it) short term, intermediate term, or long term. Generally, a short-term (or near-term) trend is less than a month. An intermediate term is up to a quarter (three months) long while a long-term can be up to a year. And to muddy the water a bit, the long-term trend may have several trends inside it (don't worry; the quiz has been cancelled).

Trendlines

This is a simple feature added to the chart to show a straight line to designate a clear path for that particular trend. The trendline simply follows the troughs in the trend to show a distinctive direction. They can also be used to identify a trend reversal or change in the opposite direction. Figure 18-4 shows a trendline.

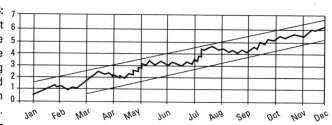

Figure 18-4:
Chart that
shows the
jagged edge
going
upward
along with
trendline.

Channels

Channel lines are lines that are added to show both the peaks and troughs of the primary trend. The top line indicates resistance (of the price movement) and the lower line indicates support. Support and resistance are important concepts in technical analysis (more about support and resistance later in this chapter). The channel can slope or point upward, downward, or go sideways. Technical traders will view the channel with interest because the assumption is that the price will continue in the direction of the channel (between resistance and support) until technical indicators signal a change. (This tells me to change to a cable channel but that's just me. Please continue reading . . .) Check out the channel in Figure 18-5.

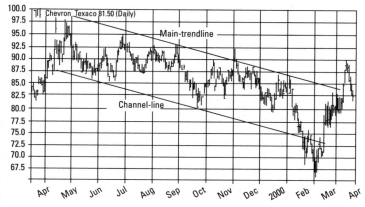

10-Apr-2000 **Open** 80.91 **High** 82.80 **Low** 80.61 **Close** 81.50 **Volume** 1.3M **Chg** -0.36 (-0.43%)

Figure 18-5:
Chart
showing
channel.

As you are seeing the asset's price moving upward, you see where the price is on both the top and bottom. The channel shows you how the price is range-bound. The continuance emphasis on trends is to help you make more profitable decisions since you are better off trading with the trend than not.

Resistance and support

The concepts of resistance and support are critical to technical analysis the way tires are to cars. When the rubber meets the road, you want to know where the price is going.

Resistance is like the proverbial glass ceiling in the market's world of price movement. As the price keeps moving up, how high can or will it go? That's the $64,000 question and technical analysts watch this closely. Breaking through resistance is considered a positive sign for the price and the expectation is definitely bullish.

Support is the lowest point or level that the price is trading at. When the price goes down and hits this level, it is expected to bounce back but what happens when it goes below the support level? It is then considered a bearish sign and technical analysts watch closely for a potential reversal and expect the price to head down.

Both the resistance and support form the trading range for the asset's price. If the price continues in this range then it is a sideways or horizontal pattern and technical analysts watch for indicators that it will break out of the pattern and then either presume an upward or downward path.

Charts

Charts are to technical analysis what pictures are to photography. You can't avoid them 'cause you're not supposed to. That's why you see more charts in this chapter than all the other chapters combined. But trust me, especially if you are serious about trading metals (or stocks or other commodities), charts and the related technical data will come in handy.

In terms of visualization and utility, the following are the four most common charts used in technical analysis.

Line charts

A chart simply shows a series of prices plotted in a graph showing how the price moved over a period of time. The period of time could be a day, week, month, year, or longer. The prices that are usually chosen for a line chart are the closing prices for those market days.

You will notice in other chapters that I prefer to use a five-year chart because I like to encourage my clients, students, and readers to be focused on the longer term because positive results can be easier to achieve.

With a line chart, you can see how the stock has progressed during the 12-month period and you can do some simple analysis. When were the peaks? How about the troughs? What were the strongest seasons for this stock's price movement?

Bar charts

Now we are getting a little fancier. Where the line chart gives you only the closing prices for each market day, the bar chart gives you the range of trading prices for each day during that chosen time period. Each trading day is a vertical line that represents the price movements and you see the asset's high, low, and closing prices.

In a bar chart, the vertical line will have two notches. The notch on the left indicates the opening price and the notch on the right indicates the closing price. If the opening price notch is higher than the closing price notch then the line would be in red to indicate that the closing price of the asset declined versus the opening price. An up day would be in black and the closing price notch would be higher than the opening price notch.

Candlestick charts

Candlestick charts have been all the rage in recent years. They are basically bar charts but with a little more complexity to them. The full name for them is Japanese candlestick charts since they originated as a form of technical analysis in the 17th century when they were trading in rice markets. Candlestick charts are too involved to adequately describe in this space so please continue your research with the resources provided at the end of this chapter.

It stands to reason that since candlestick charts provide more information in a visual form than bar charts, they can provide more guidance in trading.

Point and figure charts

A more obscure chart that chartists use is the point-and-figure chart. When you look at it, you will notice a series of X's and O's. The X's represent upward price trends and the O's represent downward price trends.

Chart Patterns

Chart patterns are very interesting and they are the graphical language of technical analysis. For technical analysts, the pattern is important since it provides a potential harbinger for what is to come. It is not 100% accurate but it is usually accurate better than 50% of the time as odds go. In the world of trading, being right over 50% of the time can be enough. Usually a proficient technician is better than that.

Head and shoulders

The head and shoulders pattern is essentially bearish. It is usually a signal that an uptrend has ended and the pattern is set to reverse and head downward. Technical analysts consider this to be one of the most reliable patterns. The pattern shows three peaks and two troughs.

The first components of this pattern are the three peaks that break down into the tall center peak (the head) and the shorter peaks (the shoulders) that are on each side of the center peak. The two troughs form the neckline.

The head and shoulders pattern tells technical analysts that the trend that just preceded this pattern basically ran out of gas. The selling pressures build up and overpower the buyers. Hence, the price starts to come down. The shoulder on the right is like a last effort for the bullish trend to regain its traction but to no avail. Keep in mind that the neckline in this pattern is the support. As support is broken, then the tendency is that it is a bearish expectation.

Reverse head and shoulders

As you can infer, this pattern is opposite to the prior chart pattern and it is essentially bullish. This pattern signals that a downtrend has ended and is set to reverse and head upward. In this pattern, you have three troughs and two peaks. The middle trough is usually the deepest one.

In this pattern, buying pressures build up and form a base to spring upward. Remember that a bullish pattern is a series of higher highs and higher lows. In the reverse head and shoulders pattern, the neckline is resistance. Once resistance is broken then the expectation is for an upward move.

Keep in mind that with this chart (as with all charts), there is not 100% reliability or guaranteed accuracy. Technical analysts don't say that the next step after a particular pattern is a certainty; it is a probability. Probable outcomes, more times than not, tend to materialize. Increasing the probability of success for more profitable decision-making (entering or exiting a trade) is the bottom-line mission of technical analysis.

Cup and handle

This pattern is generally bullish. In the pattern, the price first peaks then craters into a bowl-shaped trough (the cup) and peaks at the end of it with a small downward move (the handle) before it moves up.

This pattern basically tells the technician that the security's price took a breather to build support and then continue the bullish pattern.

Double tops and bottoms

Both the double top and the double bottom chart patterns indicate a trend reversal. The double top is essentially a bearish pattern as the price makes two attempts (the double top) to break through resistance but failing to do so. The bottom of the trough between the two peaks indicates support. However, the two failed attempts at the resistance level are more significant than the support at the trough so this pattern signals a potential downturn for that asset's price.

The double bottom is the opposite reversal pattern. It is a bullish pattern as the support level indicators are stronger than the resistance. This signals a potential upturn in the asset's price.

As a further variation, there are also triple tops and triple bottoms. These are sideways or horizontal patterns that do portend a trend reversal. Don't even think about quadruple tops and bottoms.

Triangles

A triangle is formed when the resistance line and the support line converge to form the triangle point which shows a general direction in the asset's price movement. There are three types of triangles: symmetrical, ascending, and descending.

The symmetrical triangle points sideways so this tells you it is a horizontal pattern which becomes a setup for a move upward or downward once more price movement provides a bullish or bearish indicator. The ascending triangle is a bullish pattern while the descending triangle is bearish. Of course, if you see a divergent trapezoidal and octagonal candlestick formation supported in a bowl-shaped isosceles triangle then do nothing! Just take two aspirin and try again tomorrow.

Flags and pennants

Flags and pennants are familiar chart patterns that are short-term in nature (usually not longer than a few weeks). They are continuation patterns that are formed immediately after a sharp price movement and then usually followed by a sideways price movement.

Both the flag and the pennant are similar except that the flag is triangular while the pennant is in a channel formation. Because these patterns are so short term, they're usually considered continuation patterns.

Wedges

The wedge pattern can be either a continuation or reversal pattern. It seems to be much like a symmetrical triangle but it slants (up or down) while the symmetrical triangle generally shows a sideways movement. In addition, the wedge forms over a longer period of time (typically three to six months).

Gaps

A gap in a chart is an empty space between two trading periods. This happens when there is a substantial difference in the price between those two periods. Say that in the first period the trading range is $10 to $15 and then the next trading session opens at $20. That $5 discrepancy will be a large gap on the chart between those two periods. These gaps are typically found on bar and candlestick charts. This may happen when positive (or negative) news comes out about the security or commodity in the interim and initial buying pressure causes the price jump with the subsequent period as soon as trading commences.

There are three types of gaps: breakaway, runaway, and exhaustion. The breakaway gap forms at the start of a trend and the runaway gap forms during the middle of the trend. So obviously, what happens when the trend gets tired at the end? Why, the exhaustion gap, of course! See, this stuff isn't that hard to grasp.

Moving Averages

In terms of price data, a favorite tool of the technical analyst is the moving average. A moving average is the average price of a security or commodity over a set period of time. This is done because frequently a chart shows price movements as too jumpy and haphazard so the moving average smoothes it out to show a clearer path for the price. This helps to decipher its trend. There are three types of moving averages: simple, linear, and exponential.

Simple moving averages (SMA)

The first (and most common) type of average is referred to as a simple moving average (SMA). It is calculated by simply taking the sum of all of the past closing prices over the chosen time period and dividing the result by the number of prices used in the calculation. For example, in a ten-day simple moving average, the last ten closing prices are added together and then divided by ten.

An example of the ten-day moving average

Say that the prices for the last ten trading days are (in order) $10, $11, $12, $10, $11, $13, $14, $12, $12, and $14. It is hard to derive a trend from that but a moving average can help. First you add up all the prices; in this case the total is $119. Then you take the total of $119 and divide it by ten (the total number of trading days). You get an average price of $11.90. As you do this with more and more price data (in ten-day chronological sets) you can see a trend unfolding.

Say that on the 11th day, the closing price is $15. At this point the next 10-day trading starts with $11 (this was the closing price from the second day in our first example) and ends with a new closing price for the 10th day, $15. Now when you add up this new ten-day range, you get a total of 124. Once you divide that number by ten, you get the average of $12.40 ($124 total divided by 10 days). In this brief and simple example, we see that the 10-day moving average tells us the price trend is up (from $11.90 and moving to $12.40).

Of course, you need to see a much longer string of ten-day sets to ascertain a useful ten-day moving average but you get the point. These averages can also be plotted on a graph to depict the trend to help render a trading decision. The more time periods that you graph the easier it is to see how strong (or weak) the trend is.

Moving averages are very useful in plotting out the support and resistance levels in the trend. They are very helpful in figuring all the various peaks and troughs necessary in analyzing the trend's direction.

The most common simple moving averages

Technical analysts most frequently use 10-day, 20-day, or 50-day for short-term trading. To confirm longer-term trends, they also watch the 100-day and 200-day moving averages. Of course, there are other time frames as well but these are common.

The longer period moving averages help to put the short terms in perspective so that the trader can still view the big picture. In other words, you may have an asset correct and see its price fall significantly but does it mean that a trend has reversed? If the asset is in a long-term bull market, it is common for it to violate or go below its short-term averages (such as 10-, 20-, or 50-day moving averages) temporarily. The more serious red flags start to appear when it violates the longer-term averages such as the 200-day moving average. And if it violates the ten-year moving average . . . hey . . . watch out!

Other averages

Some critics believe that the SMA is too limited in its scope and therefore not as useful as it should be. This is why more involved variants of it are also used, such as the Linear Weighted Average (LWA) and the Exponential Moving Average (EMA). It would be too involved to adequately cover these averages in this chapter. You can get more details on them through the resources at the end of the chapter. For beginners, the SMA is sufficient.

Indicators and Oscillators

An *indicator* is a mathematical calculation that can be used with the asset's price and/or volume. The end result is a value that is used to anticipate future changes in prices.

There are two types of indicators: leading and lagging. Leading indicators help you profit by attempting to forecast what prices will do next. Leading indicators provide greater rewards at the expense of increased risk. They perform best in sideways or trading markets. They work by measuring how overbought and oversold a security is.

Lagging (or trend-following) indicators are best suited to price movements that are in relatively long trends. They don't warn you of any potential changes in prices. Lagging indicators have you buy and sell in a mature trend when there is reduced risk.

Oscillators

Oscillators are indicators that are used when you are analyzing charts that have no clear trend. Moving averages and other indicators are certainly important when the trend is clear, but oscillators are more beneficial when the asset is either is in a horizontal or sideways trading pattern, or has not been able to establish a definite trend because the market is volatile and the price action is very uneven.

Relative Strength Index (RSI)

As you read earlier, the technical conditions of overbought and oversold are important to be aware of. They are good warning flags to help you time a trade, whether that means getting in or getting out of a position. The Relative Strength Index (RSI) is a convenient metric for measuring the overbought/oversold condition. Generally, the RSI quantifies the condition and gives you a number that acts like a barometer. On a reading of 0 to 100, you can see that the RSI becomes oversold at or about the 30 level and overbought at or about the 70 level.

Moving Average Convergence/Divergence (MACD)

This is a lagging indicator that shows the relationship between two moving averages of prices. The MACD is calculated by subtracting the 26-day exponential moving average (EMA) from the 12-day EMA. A nine-day EMA of the MACD, called the signal line, is then plotted on top of the MACD, which acts as a trigger for making buy and sell orders.

That's the technical definition of the MACD but don't worry if you didn't understand it on the first round. The MACD indicator is usually provided by the technical analysis software or trading service that you may use. It's fortunately not something that you have to calculate on your own.

Crossovers and divergence

A crossover is the point when the asset's price and an indicator intersect (or cross over). This is used as a signal to make a buy or sell order. Say that a stock, for example, falls past $20 per share to $19 and the 20-day moving average is $19.50. That would be a bearish crossover and it would indicate a good time to sell or risk further downside. The opposite is true as well; there are crossovers that indicate a good time to buy.

Divergence occurs when the price of an asset and an indicator (or index or other related asset) part company and head off in opposite directions.

Divergence is considered either positive or negative, both of which are signals of changes in the price trend. Positive divergence occurs when the price of a security makes a new low while a bullish indicator starts to climb upward. Negative divergence happens when the price of the security makes a new high, but bearish indicators signal the opposite and instead close lower than the previous high.

Bollinger bands

A band plots two standard deviations away from a simple moving average. The bollinger band works like a channel and moves along with the simple moving average.

Bollinger bands help the technical analyst watch out for overbought and oversold conditions. Basically, if the price moves closer to the upper band, it is an overbought condition. If the price moves closer to the lower band, it is an oversold condition.

Alas, as I get near the end of this chapter, I just know that I didn't do the topic justice due to space limitations, but as an overview, I think that it will help you. I don't think that you have to absorb every point and arcane calculation but you should get a good grounding in the basics.

Short Term versus Long Term

I hope that this chapter gives you a good taste of technical analysis without making you scratch your head. I am personally an adherent of fundamental analysis and a value investing approach. Over the long term, history tells us that the fundamentals ultimately win. This has usually been true even though in the short term, the zigs and the zags can fool you. When the fundamentals are in your favor, any short-term move against you is a buying opportunity (provided that you chose wisely from the start). But unfortunately, too many investors are not patient and they get too busy with the short-term trees to be bothered by the long-term forest. Yet that long-term forest has a lot more green, if you know what I mean. (I hope I'm not meandering here.)

In other words, a long-term investor doesn't have to bother with things such as triangles, pennants, cup-and-handles, or other paraphernalia. Long-term investors just ask questions like "Is the company making money?" or "Are financial and economic conditions still favorable for my investment?"

Now the short term is a different animal. It requires more attention and discipline. You need to monitor all the indicators to see if your investment is on track or if the signals are warning a change in course. The technicals can be very bearish one month, very bullish the next month, and then give you mixed signals the month after that. Being a proficient technician ultimately requires more monitoring, more trading, and more hedging, which in turn likely means more commissions. For more on short-term trading, turn to Chapter 16.

Remember that all this activity also means more taxes and administrative work (tax reporting and so on). After all, who do you think will pay more in taxes, someone who buys and holds for a year or longer or someone who makes the same profit but by jumping in and jumping out based on which way the technical winds are blowing? Short-term gains don't have the same favorable rates as long-term gains. Sometimes the issue is not what you make but what you keep (taxes are covered in Chapter 20).

Resources for Technical Analysis

Use the following resources to discover more information about technical analysis:

- Stock Charts (www.stockchart.com)
- Incredible Charts (www.incrediblecharts.com)
- Trading Education (www.tradingeducation.com)
- Online Trading Concepts (www.onlinetradingconcepts.com)
- *Stocks & Commodities Magazine* (www.traders.com)
- International Federation of Technical Analysts (www.ifta.org)
- Elliott Wave International (www.elliottwave.com)
- Book: *Technical Analysis For Dummies* (www.dummies.com)

Chapter 19

Following Politics and Markets

In This Chapter

▶ Being aware of how governments affect markets

▶ Understanding the manipulation controversy

▶ Profiting from interventional analysis

In the world of investing and speculating, investors consider the usual things, such as supply-and-demand factors, market analysis, and so on. But you need to be aware of political and governmental intervention as well to minimize your risk and maximize gains.

Precious Metals and Skullduggery

This is juicy stuff. Greed. Manipulation. Drama. Conflict. And that's just before breakfast! Precious metals — in this case gold and silver — do have a side filled with international intrigue. Precious metals is a topic that reaches back to the dawn of civilization and across the globe and could end up in your wallet. International economics, politics, and financial markets have much to do with helping you profit.

The controversy over market manipulation

Although I fully cover market-based schools of thought such as fundamental and technical analysis in other chapters, I don't want to leave out another center of influence: politics. Politics and government intervention have to be considered as well.

So what does all of this government and market intrigue mean for investors? How can one profit from watching this stuff unfold?

Strategic assets

Among all the investment categories, precious metals — specifically gold and silver — are in a class of investment assets referred to as *strategic assets*. Strategic assets are a special category that gains interest not only from the usual suspects — you and I and other private investors such as brokerage firms, mutual funds, and so on — but also public entities such as governments and international government agencies such as the International Monetary Fund.

Strategic assets are assets that have a high importance to the viability and survivability of a sovereign state — the country's government and economy. Assets may become strategic, and they may also lose their strategic status. Wood, for example, was probably a strategic asset to some society centuries ago since it could be used for energy (fire), constructing shelters, and making weapons (such as bows and arrows).

Examples of current strategic assets are gold, silver, oil, and government-issued entities such as a currency. Because currencies are issued by governments, currencies can easily be used as strategic assets. A particularly good example of this is the U.S. dollar.

A currency as a strategic asset

The U.S. dollar, which is the world's reserve currency, can be considered a strategic asset since it is issued by a government (our government) and held by other countries for both economic and political purposes. Many countries have dollar reserves because they sold the U.S. goods and services and of course the U.S. paid in dollars. As of the summer of 2007, China had total dollar reserves exceeding $1 trillion since it sold the U.S. so much stuff. China's plan is to use these dollars to buy up necessary commodities and other strategic assets (such as precious metals, oil, and energy).

All of these dollars held by China have a political side to them: China has threatened to sell them on the open market, which would have a harmful effect on the value of the dollar (an inflationary impact as more dollars become available). Such events are ultimately bullish for precious metals. When politics heat up between countries, it stirs up other potentially negative exchanges between countries such as trade wars and tariffs.

Some commodities can be strategic assets given market and political conditions. A good example is uranium. During the late 1970s, the United States had an energy crisis and it was involved in a nuclear arms race with what was then the Soviet Union. Since uranium was necessary for the U.S. for both economic and military purposes, it was a strategic asset.

Strategic oil

Some assets have been strategic for centuries (such as gold). Others became strategic assets since the 1950s as the world became more industrialized and interdependent (such as oil). Oil is a great example of a strategic asset and its market does ebb and flow generally with the gold market so gold investors should be aware of this market.

Oil is a very important strategic asset for obvious reasons. Until the globe can wean itself off of it, oil will grow in importance. The U.S. has been increasingly dependent on foreign sources of oil since the 1970s. For this reason, the U.S. government maintains the Strategic Petroleum Reserve (SPR), which has (as of 2006) over 700 billion barrels of oil. This is a critical backup supply because of the increasingly hostile world; the U.S. imports most of its oil from countries that are unfriendly to it or that are politically unstable. As the peak oil situation unfolds (see some information on this in Chapter 8) this situation will get more problematic.

The impact on precious metals

In 2007 and in the years to come, due to government policy (both domestic and international), currency issues (inflation and related currency matters), and the relentless growth of the world population, the conditions are ripe for a bull market in precious metals. As precious metals (along with other strategic assets) grow in demand and visibility, increasing government participation in these markets has become a fact of life.

The bottom line for investors and speculators is that strategic assets gain the attention of unusual market participants: governments and their proxies (organizations acting on behalf of governments). Investors need to be aware of movements in these markets to make better investment decisions.

The gold market manipulation controversy

Governments have a love/hate relationship with gold. On the one hand, it is indeed a treasured asset and has been for centuries. How often throughout history did that particular king or queen tell those stout-hearted explorers of the New World they wanted gold and silver to be found and brought back. I could just imagine some ruler saying to a returning explorer, "I pay for your six-month trip to bring me back gold and silver and you come back with nothing but spices?! And little bottles of shampoo?! Guards!"

The other side of gold for governments is the part that they hate: Gold competes with manmade currencies. Throughout history, gold retained acceptance as a store of value and as a medium of exchange. For smaller transactions, silver fit the bill (literally). Currencies, on the other hand, were subject to the whims and machinations of whomever was in power. The end result is that abuse of currencies (increasing or inflating their supply) was common. There have been literally thousands of currencies throughout history and across the globe that were created, inflated, and devastated. The most common act of governmental mismanagement that impacted an entire society has been the inflating of a currency right into oblivion. It's incredible how often it has happened.

Why? Think about it. If you were a ruler and you can make your central bank create money at will by simply printing (or today, doing it electronically) money and no one can stop you, would you? Maybe not you, but how about the ruler after you? Ultimately, it gets done (monetary inflation). When people hold currency that is losing value (remember that monetary inflation begets price inflation), they'll seek alternatives, such as gold and silver.

This is why governments frequently seek ways to suppress the price of gold. Gold is more than an enduring asset and monetary metal. In modern times it acts like a barometer of problematic economic ills (such as inflation). When gold goes up in value, that becomes a signal to the public that something is wrong. History bears this out as well.

During the late 1960s, some governments with a few favored private financial institutions conspired to suppress the price of gold. This was the scandalous "London Gold Pool" affair that ultimately failed. During this decade, the issue of gold market manipulation does take on credibility as a growing body of research and market data suggests that it is occurring (compiled by organizations such as the Gold Anti-Trust Action committee, which you can check out at www.gata.org).

The Plunge Protection Team (PPT)

In the wake of the 1987 stock market crash, the government decided to do something that it believed should be done to protect investors from what seemed to be shocking gyrations in the market. As a result, a presidential directive (Executive order 12631) was issued in early 1988 which created the "Working Group on Financial Markets." It was composed of the U.S. Secretary of the Treasury, the chairman of the Federal Reserve (the central bank of the U.S.), the chairmen of the Securities and Exchange Commission (SEC), the Commodities Futures Trading Commission (CFTC), the heads of some large Wall Street firms, and a few others. They became a task force that would use resources and funds provided by the federal government to intervene when the market is experiencing a particularly bad day. This committee was nick-named the Plunge Protection Team (PPT).

The PPT did its work in a low-profile way so that the general public was not aware of its activity. Its activity became more noticeable after market-shaking events such as the horrific terrorist strikes on September 11, 2001. In due course, more congressional testimony and publicly available data (such as minutes from meetings) slowly became available. It became evident that some intervention did take place, either to keep the stock market from falling or to keep some strategic assets from rising. The most common way that gold's price was kept from rising too far was by efforts from central banks such as the Federal Reserve. The retired Fed chairman, Alan Greenspan, did admit that some intervention was taking place.

There were some allegations that oil, gold, and other strategic assets were targeted to be brought down during the summer and fall of 2006 (by the government's proxy brokerage firms through massive shorting techniques in the futures market) to affect outcomes in the November 2006 elections, but that is purely speculation at this point (until more data and evidence are made public). Stay tuned.

The profit in market meddling

If you owned an asset and you saw its price plummet, it would be disconcerting for you. If you saw it plummet because someone forced its price down, that would be infuriating. But if you are aware of how it works, it can actually help you profit.

The price of an asset (such as a stock or commodity) can go up or down due to one (or both) of two factors: natural and artificial forces.

Natural forces

Natural forces means the voluntary buying and selling of market participants (a free market). *Free markets* mean voluntary participation by willing buyers and willing sellers. In a proper market, the government's most desirable role is a limited one; it acts like a referee. Technically (and morally) it should not be a participant. This is supply and demand at its finest, and it has been the most efficient human mechanism for discovering the market value and price of goods, services, and assets. Those natural forces are folks like you and me and umpteen consumers, producers, investors, and speculators. Natural forces through the age-old formula of supply-and-demand result in natural prices which are in fact an important concept for investors.

Artificial forces

Artificial forces means involuntary actions by entities that have an undue ability to force an asset's price up or down. This is typically embodied through governmental actions ranging from (but not limited to) taxes and customs tariffs to market regulations and price controls. The key point to remember is that government is a coercive instrument no matter how beneficial the objective or result may be (either real or perceived). Economists such as Ludwig von Mises (www.mises.org), Henry Hazlitt, and many others accurately pointed out the results of this intervention.

Government intervention can affect the price of an asset in the short term, but ultimately market forces bring the price of the asset to its natural price over a longer period of time. If the natural price of an asset is, say, $50 and government intervention has forced its price down to $25, then this coercion tends to have the same impact as you forcing a spring down; ultimately the spring whipsaws back up and typically to a higher point than its natural level (at least temporarily). In the example of the $50 asset being forced down to $25, the end result will probably be that it will go to, say, $75 before market forces bring it to a natural level.

Artificially forcing a price down, for example, can be done through governmental price controls. In other words, why not just pass a law decreeing that some asset (or product or service) is fixed (or legally forced) to be at a mandated price? Keep in mind that when an asset's price is artificially low, then demand for it will be artificially stimulated while supply becomes artificially suppressed. If producers make a profit at a natural price of $50 and the law states that the price must be $25 then the natural result will be that producers will make less while consumers are encouraged (by the artificially low price) to buy or consume more. Shortages will then result and will persist until natural forces of supply and demand operate again.

Some unlikely buying and selling opportunities

In other words, government intervention (such as price controls or market manipulation schemes) ultimately backfire. If gold, silver, and other finite, valuable assets have a price that is artificially lower then what you have is a classic buying opportunity. Let's summarize. For informed investors, the following points offer opportunities:

✔ When an asset's price is artificially suppressed below its natural price, the asset becomes *undervalued* and hence becomes a buying opportunity.

✔ When an asset's price is artificially raised above its natural price, the asset becomes *overvalued* and hence becomes a selling opportunity.

I like the way that one market analyst put it (I wish I thought of it). He called these government forces *interventionals*. So when he analyzes an investment, he looks at three things: the fundamentals, the technicals, and the interventionals. In other words, he uses fundamental analysis, technical analysis, and interventional analysis to factor in government-related intervention.

A good example of this was China and how it played the copper market in the middle of this decade. China (through its intermediaries) went into the futures market to aggressively short-sell copper futures. This heavy selling forced the price of copper down below its market price. It then proceeded to buy physical copper by the boatload. Immediately after this slight-of-hand intervention, the price of copper skyrocketed. At that time, astute investors saw that copper's price was lower, yet their fundamental analysis suggested that copper was very bullish. Anyone who bought (or went long) copper futures or copper-related investments (such as copper mining stocks) would have enjoyed some excellent gains.

Yet another take on this singular area is a variation on it called inter-market analysis (IMA), which tries to figure out how one market can and will affect another market; that includes government intervention. In this approach, one doesn't analyze government actions directly, but indirectly. If, for example, a government entity is participating in the futures market, then IMA would take into consideration the futures market activity and how that would affect related markets (such as in the physical market).

Either way, it behooves the serious investor to be aware of the interventionals. This is more important than ever since governments across the globe are vying more and more over natural resources. Among them, strategic assets such as precious metals.

The metals more affected by the interventionals are gold and silver. Resources that help you understand this are listed at the end of this chapter.

Precious Metals and Geo-Politics

In spite of the effects of government intervention in the world of strategic assets, the global market has become more competitive so it is highly likely that supply and demand on a global scale will win out over any intervention that may be affecting us in American markets. There is some interesting jockeying for position internationally.

While central banks (technically national governmental banks) such as the Fed and central banks in Europe have been selling off their gold reserves (which either unwittingly or purposely has artificially driven down the price of gold), other central banks (such as in China, Russia, and the Middle East) have been buying gold. In the age of increasing inflation, this is disturbing. Isn't more gold in the national treasury a good thing especially since currencies have collectively been shrinking in value due to excessive creation?

The unfolding worldwide supply and demand situation is a bullish harbinger for precious metals. It was recently reported by the respected international metals consulting firm, the CPM Group (www.cpmgroup.com), in their annual publication, *The CPM Gold Yearbook 2007* a very interesting fact. The year 2006 was the first year in world history that private investors (individuals and private organizations) owned more gold, cumulatively, than the world's central banks. Go get 'em!

Couple this with the fact that demand for precious metals in China, India, and the Middle East is projected to increase and you have a very bright and shiny outlook for gold, silver, and other precious metals.

Resources on Politics and Markets

I think that politics and markets can be fascinating. Who needs sports or soap operas when the markets have plenty of real-life drama and excitement? To find out more about the interplay of governments, financial markets, and precious metals, here are some places to start:

- ✔ Le Metro Pole Café (www.lemetropolecafe.com)

- ✔ Gold Eagle (www.gold-eagle.com)

- ✔ Safe Haven (www.safehave.com)

- ✔ Financial Sense (www.financialsense.com)

- ✔ For Fed watchers, visit the Federal Reserve Web site (www.federal reserve.gov)

- ✔ To watch out for political developments, new laws, and regulations, check out the excellent search engine covering new and old laws called Thomas. It can be accessed from the home page at the Library of Congress (www.loc.gov).

Chapter 20

Dealing with Taxes

. . . .

. . . .

Complicated. Expensive. Annoying. Unfair. Of course, taxes also have their negative side, too. This is life and we deal with it. And look . . . if you're paying taxes, that means you're making money so there is a bright side. The other side of tax obligations is, of course, tax benefits. In this chapter, I explore both as they apply to the world of precious metals and related investments.

Keep in mind that tax laws change quickly, too quickly. If this chapter is to be relevant and accurate (as relevant and as accurate can be given relentlessly changing tax laws) then it will be necessary to not get too specific. So you should check and double-check with your tax advisor.

Taxable Activity

First things first. Before I discuss any tax liability from any investment transactions, you first need to find out if there is any liability to begin with. Keep in mind that your investment and speculative activity will generally take place in one of two types of accounts: a regular account with regular activity and a tax-sheltered account, such as an individual retirement account (IRA) or other retirement account (such as a 401k plan). I cover both of these accounts in the sections that follow.

You should be aware of whom you're paying tax to. In this chapter, although the major concern is paying your federal taxes, you will probably also have to deal with state and local taxes.

The regular account

The regular account is one almost all of us have. There are no special rules except to pay tax on the following transactions:

- ✔ **Dividends:** Dividends are payouts by a company to its shareholders. In a regular account, dividends are taxable.

- ✔ **Interest:** Interest is paid by debtors to creditors. You can receive interest from holding bank accounts, corporate bonds, and United States Treasury securities (such as treasury bills and savings bonds). Municipal bonds are generally tax-free, but find out about your own state's tax laws.

- ✔ **Capital gains:** A capital gain is any gain you generate when you sell a security or asset. If you bought something for $10 and you sell it for $14, then your taxable capital gain is $4. Remember that for regular investors, you pay the capital gains tax on gains that are realized.

Those are the basic taxable transactions you will need to deal with. If your regular account is with a stock brokerage firm, it will provide you with a year-end 1099 and you should keep all statements. In addition, brokerage firms usually let their customers download trading data from their Web sites, which makes it easy to calculate gains and losses.

Table 20-1 shows you the types of precious metals assets covered in this book and the potential gains you may have. This table gives you a snapshot of what to watch out for and what could potentially be taxable with precious metals. Check the special rules for gold and silver later in this chapter.

Table 20-1	Assets and Gains		
Asset	*Typical types of gains**	*Found in Chapter*	*Comments*
Bullion coins and bars	Capital gains	10	Taxed as collectible**
Numismatic coins	Capital gains	11	Taxed as collectible**
Junior mining stocks	Capital gains	12	The broker sends a 1099
Major mining stocks	Capital gains and maybe some dividends	12	The broker sends a 1099
ETFs	Capital gains and maybe some dividends	13	Taxed as collectible **

Asset	Typical types of gains*	Found in chapter	Comments
Mutual funds	Capital gains and maybe dividends and interest	13	The fund will send you a 1099
Futures	Capital gains	14	The broker sends a 1099
Options on stocks and ETFs	Capital gains	15	The broker sends a 1099
Options on futures	Capital gains	14 and 15	The broker sends a 1099

*In this table I mention only taxable gains (such as capital gains). Of course there are capital losses and they are generally deductible. For more information, see the section on capital gains and losses later in this chapter.
**Some types of precious metals investments are classified as "collectibles" and therefore get different tax treatment (see the section "Special Tax Considerations" later in this chapter). The treatment is actually odd and unfair but it is something to be aware of when the time comes to sell your precious metals investment.s

One of the ways around the current taxation of your gains in a regular account is to be aware of the benefits in the second basic type of account, the retirement account.

The individual retirement account

Although there are different types of retirement accounts, I'll just comment about precious metals–related investments inside a typical retirement account. The most common types are Roth IRAs and traditional IRAs. In addition, many firms allow employees (past or present) to rollover the proceeds from a 401k plan to an IRA (called a "rollover IRA" of course!). For simplicity, I just refer to each of these various accounts as an IRA.

The biggest benefit of the IRA is that any gains generated in the account are shielded from current taxes. Except for physical bullion, most forms of precious metal investments (especially paper ones such as stocks and ETFs) can be inside an IRA. Inside the IRA, any dividends and interest can be reinvested and are sheltered from current taxes. A major criteria in the IRA is that you don't withdraw any money until retirement time (you can start in the tax year when you turn 59½ years old), although there are some allowable exceptions.

Because capital gains are the primary reason for getting into precious metals, buying and selling for gain can take place inside the IRA with no tax consequence. You won't have to worry about taxes until you start removing money from the account.

By the way, if you want to find a way to place gold and silver bullion into an IRA, get more details in Chapter 10. The IRA is also a way to remove the pesky unfair tax treatment of gold and silver bullion (as collectibles).

Reporting your activity

As a U.S. citizen, you report your annual taxes using Form 1040. Investment income such as dividends and interest are reported on Schedule B. Capital gains and losses are reported on Schedule D. Deductible investment expenses are taken on Schedule A. These schedules cover most of the reporting that most individuals are required to do. As always, check with your tax advisor if you have other transactions and issues in your particular situation.

Capital Gains and Losses

The most common way to profit from precious metals–related investments is through capital gains. The most basic point about capital gains is that they're taxable (see exception above regarding retirement accounts) and the tax rate is different based on how long you hold the asset. Capital losses are generally tax-deductible.

To keep it straight, there are two basic time frames for capital gains tax rates. There are short-term gains (one year or less) and long-term gains (over one year). Short-term capital gains are taxed as ordinary income, which is the same basic tax rate as income that you earn at your job or business. Ordinary income tax rates are considered the highest individual taxes and because we have a progressive tax system, the more income you earn, the higher the taxes you pay. Higher taxes refers to paying more in taxes as well as higher tax rates (as in, you've made more money that bumps you into a higher tax bracket). So if you're not careful it ends up being a double whammy. Check with your tax advisor.

Technically, a loss is not deductible in the same sense as when you pay a tax-deductible expense. A loss is deductible in the sense that you can apply it (offset it) against other income. In other words, if you have income from your job of $30,000 and you have a loss from the sale of stock of $2,000 then your gross income for tax purposes would be $28,000.

Generally, any capital losses you have must first offset capital gains before they can be used against other income on your Form 1040 individual tax return. After that, you can use up to a maximum of only $3,000 of capital losses per tax return per year. Any amount above that is treated as "carry-forward" losses that can be taken in future years. For more information about gains and losses, check out IRS publication 544.

Tax-Deductible Activity

After touching on the ugly side of taxes (paying them) we can then see the beauty of reducing their payment (tax deductions). There is no need to go nuts here and list all of the possible deductions but there are many.

As a general rule of thumb, the IRS allows deductibility of investment expenses as long as they are ordinary and necessary expenses paid or incurred to either produce or collect income or manage property held for producing income. The expenses must be directly related to the income or income-producing property and the income must be taxable to you. For most investors, the following deductions generally fit that criteria and can be reported on Schedule A:

- Attorney or accounting fees
- Investment counsel and advice (which includes advisory services)
- Fees to collect income
- Safe deposit box rental fees
- Investment interest

By the way, commissions you pay when you are buying and selling securities are not deducted directly. Instead, they become part of the cost basis of the sale. For example, if you buy a security for $4,000 and the commission is $20, the actual cost basis (when you figure your gain/loss at the sale later on) is $4,020. If you sold that same security later on and sold it for $5,000 (with another commission of $20) then the sale amount would be $4,980. The gain would then be $960 ($4,980 sale amount less the cost of $4,020).

Keep in mind that the IRS disallows tax deductibility of some items such as travel expenses to a stockholders' meeting or the cost of an investment-related seminar. IRS Publication 550 goes into greater detail.

Special Tax Considerations

For precious metals, you will probably bump into some tax concerns that are especially related to your pursuits. In the sections that follow, I describe a couple of these concerns that you should consider.

Gold and silver as collectibles

Gold and silver are considered capital assets (unless you are a dealer, in which case they are effectively merchandise and the income is ordinary business income). So far, so good. However, they are considered in the same class as collectibles such as stamps or art. Why does this matter? Collectibles are taxed at the maximum 28% federal rate even if you hold them for more than a year. The rule applies to both bullion and numismatic coins and bars. If you hold the physical metal for less than a year, then the gain would be taxed as ordinary income up to the maximum 35% federal rate. That is astonishing and disappointing but that is the law. This rule also applies to bullion-specific ETFs such as the Gold (symbol: GLD) and Silver (symbol: SLV) ETFs.

The tax rates hit you when the gains are realized. If you hold these properties and they have appreciated, there is no tax consequence until the actual sale.

Fortunately, the rule generally does not affect other precious metals such as mining stocks, options, futures, and metals-related mutual funds. Hopefully, future tax laws will correct this unfairness.

Tax rules for traders

If you are actively trading futures contracts or options on futures contracts, check with your tax advisor about how to proceed. If you are the occasional trader or speculator, the odds are that the regular tax rules would apply (not that the regular rules are so good to begin with but read on). But what if you're an active trader?

You may qualify as a professional trader conducting a business. It could mean more tax deductions for you (hey . . . sounds good), but your trading then takes on a different characteristic. It is now a business activity and the trading you do gets treated much differently. Welcome to the world of mark-to-market accounting.

Mark-to-market (MTM) is an accounting procedure by which assets are *marked,* or recorded, at their current market value, which may be higher or lower than their purchase price or book value.

Part of the benefit of mark-to-market accounting is that you don't have the same tax constraints as investors. Investors, for example, have a capital loss limit of $3,000 but not with MTM because they're considered a trader doing business and not an investor doing personal transactions. MTM accounting can be complicated and you definitely need professional tax assistance with it.

Tax Resources to Keep You Up-to-Date

Yeah . . . it's all pretty heady stuff. You may end up needing to make a fortune with your precious metals just to have the money to pay your taxes. Fortunately, it's not that bad. And besides, there are countries with steeper tax rates (like Canada and France) so you may want to consider moving to lower-tax places (such as Ireland and the Ukraine). But you can do just fine here in the good 'ol USA by doing a little homework. The old adage fits: The more you know about taxes, the more money you end up keeping. Keep informed and hopefully keep more money by checking out the resources I discuss in the next sections.

The IRS . . . of course!

What better source on taxes than the IRS itself? You pay them the big bucks so you might as well get the free assistance you paid for. You can get the following publications from the IRS by either calling 800-TAX-FORM (800-829-3676) or by downloading them at `www.irs.gov`. The following publications go hand in hand with investing in precious metals:

- **Publication 17 Your Federal Income Tax:** If you file form 1040 (who doesn't?) then this extensive guide will be very helpful.

- **Publication 550 Investment Income and Expenses:** This publication is tied directly to this chapter so cozy up with it tonight.

- **Publication 590 Individual Retirement Arrangements:** This publication covers IRAs in-depth and is worth a peek.

- **Publication 544 Sales and other Dispositions of Assets:** This publication goes into greater detail about capital gains and losses (investments and otherwise).

If you have any tax questions and need to speak with someone, call the IRS during business hours at 800-TAX-1040 (800-829-1040). They have extended hours during the tax season.

Helpful tax Web sites

You can check out the following Web sites for additional tax help:

- Tax Topics (`www.taxtopics.net`)
- Tax Sites (`www.taxsites.com`)

- ✔ Tax Cut Central (www.taxcutcentral.com)
- ✔ National Association of Enrolled Agents (www.naea.org)
- ✔ Fair Mark (www.fairmark.com)

An ounce of prevention . . .

Perhaps the best recommendation I can give you is the last thing I want to express. Plan your future taxes. You heard that right. The best way to minimize the tax bite is to start tax planning before the year is up. Most taxpayers start figuring out their taxes after the tax year is done. In other words, millions of taxpayers try to figure out how to lower their tax bills (or increase their tax refunds) for the tax year 2007 in the year 2008. With 2007 over with, you can't change a darn thing; you can't change history! You can only report what happened and spend time finding deductions and credits for what you already did. Too late!

The best way to prepare your taxes (for, say, 2007) is to start long before 2007 ends. Keep in mind that when you do your taxable transactions, the timing is within your power. You can control when you transact, which means you have the power to defer, and you have the power to accelerate transactions (like paying tax-deductible expenses). In the world of taxes, timing means a lot. Selling securities is a perfect example: If you sold your losing positions in December and you sold your winning positions the following month (in January, which is part of the subsequent tax year), you defer taxes on your winning positions until the next tax year. Keep good records of your transactions and have a tax planning session with your tax professional long before the tax year ends. That way you will have a profitable jump on your taxes when you can do something about them.

Part V:
The Part of Tens

By Rich Tennant

"So, what kind of metals did you want to invest in? Precious, cute, or just adorable?"

In this part . . .

You can't have a *For Dummies* guide without the Part of Tens. The lists in this part keep you on the right track with your precious metals activity. You can find out more about mining stocks, get some helpful guidelines for investing and trading, and even discover some great ways to limit your risk.

Chapter 21

Ten (Nearly) Reminders about Mining Stocks

In This Chapter

▶ Knowing what to look for

▶ Checking on important financial details

▶ Understanding the effects of regulatory and political environments

*1*n stock investing, you learn to look at the company from various perspectives. One analyzes the company's products/services, management team, financial strength, and so on. Beyond the usual, you check out those things that are indigent to their market or industry. In that way, every group of stocks has its own quirks and specialties. Mining stocks are no different.

The Company's Management

The people who run an enterprise make all the difference in the world. Good management can take mediocre properties to a profitable end result. Mediocre or incompetent management is better off looking for loose change with a snazzy metal detector over at the local public beach. Odds are they'll find base metals instead of precious metals (aluminum soda cans do have value).

Basically you want the company led by people who have experience, expertise, and a track record. An easy way to do this is to see their resumés (usually posted at the company Web site). Here are some things that are a must:

✔ **Experience:** Do the executives on the management team have experience with well-known companies in the mining industry? You should be able to see plenty of references like "The President, Abner Kaputski, has had 47 years of experience" at large, recognizable mining companies such as Newmont Mining or Goldcorp.

✔ **Expertise:** Besides management experience, they need to have experience in areas directly related to precious metals such as geology and metallurgy. Having degrees from renown schools is a big plus.

✔ **Track record:** Have these executives had successful projects such as finding properties with proven reserves? Have their projects been successfully converted into profitability for the company?

✔ **Other experience:** Besides precious metals–related experience, other experience is valuable. If, for example, the company has operations in Australia, how much experience do they have with Australia in terms of markets and governmental issues (such as Australian mining regulations)?

Having a management team that is highly competent and experienced makes everything else happen. Their experience, expertise, and track record should not be a guessing game; it should be verifiable. For more information on mining stocks, see Chapter 12.

Financing

The whole point of mining is to extract some valuable stuff from the ground (gold, silver, and so on) and sell it to make a profit. This is a simple point but it is not easy to do. It requires a lot of money to pay for everything from mining equipment, to hiring geological staff, regulatory costs, and much more. A mining project could take years before you see a dime of profit. So financing is a critical issue. There are some questions that need good answers:

✔ Does the firm have ample lines of credit to hold it through times when profits are not rolling in?

✔ Does the firm have cash in the bank to cover the day-to-day expenses until the metals are extracted and sold?

✔ Does the firm have joint ventures with other firms that have financial clout (such as a major mining company)?

Mining can be a difficult and daunting task that can take a long time. Much of the time is tied to all sorts of regulatory hurdles that must be met before any substantive progress can be made. Until that progress materializes, the company's bills need to be paid. Financing in the form of cash on hand or credit (or both) will be instrumental for the company's viability.

Earnings and Cash Flow

Consistent earnings is not an issue regarding junior exploration and drilling companies because earnings are at best inconsistent. The point about consistent earnings relates to the majors — those large mining firms that are already extracting metals and have established sales. For those investors who are seeking conservative growth, major mining firms with consistent earnings are a reasonable objective that will provide a catalyst to keep pushing the company's stock price upward as the bull market unfolds.

How consistent should the earnings be? Three years or more is a good benchmark. In addition, each year's profits should be greater than the year before. Many analysts will also point out the cash flow is another important factor and can sometimes be more important than reported earnings. Cash flow means funds coming into the company, which may or may not include earnings. Cash flow includes money coming in from the sale of metals/minerals but also from other things, such as the sale of assets or revenue from rents and property leasing. The point is that cash must be flowing in to cover the day-to-day expenses until potentially profitable projects come to fruition.

The company's publicly available financial statements can provide this information.

Balance Sheet Strengths

The balance sheet gives us a snapshot of where the company stands in regards to its net worth (or stockholders' equity). The balance sheet reflects the equation: total assets minus total liabilities equals net worth. The balance sheet is key to the company's value, both real and perceived.

This is especially true if the company is a junior mining/exploratory firm because it may not yet have revenue because they are occupied with getting the metals out of the ground first. For these companies, the following strengths in their balance sheet are a must:

- **Proven reserves:** If the company is sitting on property that has lots of metals/minerals, this is critical to their success.
- **Cash on hand:** Do they have money to pay the bills until their drilling/exploring bears fruit?

These numbers are easy to find at the company's Web site or in its 10-K reports which are found at www.sec.gov (use the stock symbol and find these documents through the SEC's EDGAR database at their Web site).

Regulatory Environment

The regulatory environment is the framework inside which the company has to operate. The regulatory environment is created and maintained by the government. Just because a company is allowed to operate in an area, that doesn't mean that it can operate successfully. A good example of this is in juxtaposing California with its neighboring state, Nevada.

California has a stringent regulatory environment that can effectively be hostile to developing mining properties. I know of one company there that is still waiting (after ten years!) for approval to develop a mining property in that state. Nevada, on the other hand, is the most mining-friendly state in the union. Nevada, to its credit, streamlined the regulatory process to make it easier to mine while still adhering to environmental standards.

A mining enterprise always has to face regulatory hurdles that can be present at many levels of government: federal, state, and local. Some hurdles are good, of course, to address environmental concerns. However, sometimes regulations can cross the line from being necessary to just being a barrier that ultimately means more costs and delays for the company and its customers.

Keep in mind that regulations go beyond the U.S. It also means the countries across the globe that allow mining. Just because a country is "politically friendly" that doesn't always mean that it has a conducive regulatory environment. Companies generally send off press releases regarding what approval they have achieved so check their Web sites for details.

Hedging and Forward Sales

Sometimes a good company with good management can make a miscalculation when it comes to the practice of hedging, which can also be called *forward sales*. Hedging is something that could harm a company that does it in the wrong market conditions.

Hedging refers to an industry practice of selling next year's mine production and locking in this year's price. Hedging is a way to receive predictable income for the coming year, and it can be a good practice if what you are selling has a price that is stable or declining during that year. However, it can be a real problem if what you are selling is going up in price.

Say a company called Ironic Gold Mining Company (IGM) has a hedging program and they are committed to selling one million ounces of gold (it can be other metals, too) next year at this year's price of $600 per ounce. If next year's price of gold is $600 or less, then it was a good decision. But what if gold is in a bull market and the price is higher next year? If gold goes to $650,

$700, or more, then the company won't able to make the extra profit. IGM is stuck with the commitment to deliver that gold at $600 per ounce no matter how high the market price of gold goes.

It indeed becomes an ironic situation; a gold-mining company losing money in a gold bull market! It's like a fish drowning in water. The bottom line is that hedging for a mining company is fine when metals prices are flat or declining (bear market), but it is a bad practice in a bull market. Check the company's annual report (or Web site) to see if it engages in hedging and to what extent.

Valuable Projects or Properties

The more properties that a company has that are producing properties and/or properties with proven reserves, the more valuable that company will be, especially in a precious metals bull market. Again, check out the company's Web site and/or annual report for details.

A good analytical tool is to find out how many ounces of metal they have and multiply it by the market price. Taking it a step further, many analysts will look at the total shares outstanding and the total value of its metals to come out to an asset or "book value" per share. I know of one silver company that is at $37 per share but when you tally up the value of its total reserves, the book value is at over $200 per share! In a precious metals bull market, that company would be very valuable and could even be a takeover candidate by other, larger mining companies that need reserves.

Extraction on Cost Per Ounce

This point gets to the heart of profitability. The more a company can be performing cost-effectively and economically, the more profit it can make. The more profit it can make, the better the prospects are for its stock price.

The extraction cost on a cost per ounce basis is not difficult to find out and usually the work is already done for you. If you have a gold-mining company that can successfully mine gold at a cost of $200 per ounce and the price of gold is over $600 an ounce, is it profitable? At a $400 gross profit per ounce, I can concur with a "yee-haa!" which is Croatian for "nice profit." Yes, mining costs can and do go up. But . . . the price of what you are mining can and does go up.

The wider the spread between the company's cost of mining and the price of what it mines (in this case, the price of gold) the more profitable the company can be (and the happier the shareholders are).

Political Considerations

Governments can be hostile to gold in general and to mining companies that own and operate properties in that government's jurisdiction. Government can enact high taxes and fees and/or tough restrictions. They can also simply seize the property or otherwise harm the company. A hostile government can break a company so this issue is an important consideration for investors. For more details on this, check out Chapter 19.

The bottom line here is to avoid companies with exposure to problematic governments (as of August 2007, good examples are Russia and Venezuela). You are better off sticking to companies that operate on friendly terrain (good examples are the United States, Canada, and Australia).

Chapter 22

Ten Rules for Metals Investors

In This Chapter

▶ Diversifying for success

▶ Using different forms of precious metals

▶ Understanding changes in the markets and adjusting appropriately

*P*recious metals were an important part of a well-balanced portfolio in the 1970s, and they play an even more important part during this (and the next) decade. But that doesn't make investing in precious metals a "no-brainer." You need to keep some things in mind to maximize the gains and minimize the risks.

Diversifying Your Vehicles

Although society has forced families into two-car environments, I'm talking about making sure you have an SUV and a hybrid in your driveway. I'm referring to diversifying in precious metals by utilizing both paper and physical vehicles. By the way, just because SUVs and hybrid cars have precious and base metals, they are depreciating assets!

You have heard about diversification being a key factor in the long-term success and viability of your portfolio for what seems like forever. But diversification isn't just a good thing for the overall portfolio; it is also important for the particular segments inside your portfolio. Because precious metals come in different forms and vehicles, each of these has its own merits and drawbacks. All the prior chapters cover these points exhaustively (believe me; I'm pooped).

Diversification can be a double-edged sword. It can minimize risk, but it can minimize gains as well. However, if the overall precious metals market is performing well, then your diversified portfolio of precious metals investments and related vehicles should also perform well. At a bare minimum, you should have a portfolio with 10% precious metals investments in both physical and paper forms. If you had 50% of the precious metals portion of your portfolio in physical and 50% in paper form (such as stocks and ETFs), that would be a simple and effective diversification. That could be quite appropriate for conservative investors.

For those who consider themselves to be aggressive investors or speculators, a different diversification could fit. A sample model precious metals portfolio could be 30% physical, 40% conservative paper (such as larger mining company stock and ETFs) and the remaining 30% could (should) be a mix of junior mining stocks and options.

For those investors who are seeking income, consider the conservative portfolio coupled with writing options such as covered calls to generate income. (Chapter 15 has more on writing covered call options.)

Having Some Bullion Coins

Having physical precious metals is a good part of the foundation of your portfolio. Although I like collectible numismatic coins (as covered extensively in Chapter 11), it may not be suitable for beginners because of all the issues involved that one must be aware of (such as grading, mintage, dealer markups, and so on), so bullion coins make a lot of sense (cents?).

The best ones to concentrate on are gold and silver bullion coins and issues that have a large market to make it easy to buy and/or sell them. The U.S. Mint issues both gold and silver eagles, and they can be affordable. The gold eagles come in 1 ounce, ½ ounce, ¼ ounce and even ⅒ ounce sizes, which make owning physical gold convenient. A silver eagle is 1 ounce of silver and also affordable. Many investors also like to acquire bags of "junk silver" which also have a large and fluid market. More details on bullion coins are in Chapter 10.

Part of bullion's appeal is its enduring value and that you don't have to worry about a third party's performance. An example is a mining stock. You could have say, a gold-mining company's stock price plunge due to factors, such as company mismanagement or external factors, such as political risk. A variety of factors could plausibly make a gold-mining stock go bankrupt even when gold's price is rising. Owning physical gold or silver in bullion form gives you a measure of safety when there are so many things that could go wrong in the world, precious metals–related or not.

The bottom line is that bullion coins are a reasonable, convenient and relatively low-risk way to participate in a long-term bull market in precious metals. Every serious investor (and a few light-hearted speculators) should consider bullion.

Limiting Your Exposure

You may think that this is advice for you sun bathers out there, but it's another wrinkle to iron out in your portfolio. Is there any single investment that occupies too large a percentage of your total financial picture? Yes, if it does well you would look like a genius and you could go shopping for that yacht that you always dreamed of. However, more times than not, it can have a negative impact. Either reduce how much you have or watch it like a hawk by using a disciplined approach (such as trailing stops, which are detailed in Chapter 16).

Watching the Markets That Affect Precious Metals

Precious metals don't operate in a vacuum. They do get affected, positively or negatively, by other markets that may not seem related to precious metals. The same way that aspirin and bear markets are connected (think about it), so precious metals are affected by:

- ✔ Currency markets — if national currencies (such as the dollar or yen) are strong, this generally translates into weaker or lower prices for metals.

- ✔ News about rising prices at the grocery store or the gasoline pump alert people to price inflation and hence will have a positive impact on gold and silver.

- ✔ Reports of metal supply shortages in countries with extensive mining industries would have an upward effect on precious metals.

All investors and speculators should make it second nature in their decision-making process to look at the cause-and-effect aspects of the economy and financial markets. There is in fact an entire school of thought on this very idea: inter-market analysis. It makes a lot of sense because markets and economies across the world are becoming more interdependent.

For precious metals, the most obvious markets to watch are

- ✔ **Currencies:** Especially the dollar and the euro.

- ✔ **Industry:** Any industry that produces or uses a lot of metal in its processes (both in the U.S. and foreign markets).

- ✔ **Commodities:** Food and energy are sensitive to inflation and supply-and-demand factors.

- ✔ **Money supply:** The growth of the money supply is a leading factor in the precious metals market.

Using Options to Boost Performance

Options are a fantastic tool in the investors' wealth-building arsenal. Just about any portfolio, aggressive or conservative, could benefit from some options strategy.

Whether you use options to generate income (covered call-writing) or as insurance against a drop in your stock's price (protective put), look into options. For more information, check out Chapter 15.

Adding Alternatives

Exchange-traded funds (ETFs) and mutual funds (MFs) are ideal additions to any portfolio. Because there are some excellent precious metals (and mining stocks) ETFs and MFs, they make a convenient way to automatically add a diversified investment to your portfolio.

I could write a whole chapter on these great securities. Wait a minute... I did! Check out Chapter 13 and be glad I remembered.

Adjusting along the Way

No investment strategy is set in stone. Set-it-and-forget-it is a fine anthem for that gizmo advertised on late-night TV, but markets are constantly moving and changing. Companies are constantly changing. The political and economic environments are constantly changing. As you have investments that have performed well and have been with you over the years, modest or major adjustment is prudent. Pruning and rebalancing the portfolio is not done for cosmetic purposes; you need your investments to keep growing. Bull and bear markets go a long way but they do end at some point.

You don't have to time it exactly but as you read the news and industry reports, you start to get clues and warnings about what's happening. Sometimes those clues come from the public and not the usual suspects in the financial media. When the whole world was crowing about getting rich with Internet stocks in 1999, prudent investors started adjusting their portfolios.

Start either paring back some of your holdings (or at least put trailing stops on) when you see lots of media attention about precious metals and how average folks are getting rich. Make those adjustments. It's better to be a year too early than a day too late.

Understanding the Difference between a Correction and a Bear Market

In the summer of 2006, precious metals, energy, and other commodities experienced a *correction* or a temporary pull-back in their prices. Gee, why say correction? It sure as heck doesn't feel correct (it feels pretty incorrect actually). Anyway, a correction is a temporary, normal, and healthy part of a bull market. Sure it might make you twinge when you see your investment go down 10% or 20%. It is not unusual to see precious metals corrections go down an unsettling 30%. What are you going to do?

First, understand the difference between a correction and a bear market. It is like the difference between fainting and dropping dead. In a healthy, long-term bull market, frequent corrections are part of the process. Prices drop temporarily for various reasons such as profit taking and similar events that are not evidence of a fundamental change in the prospects of that asset or industry.

Take gold for example. If you saw a chart spanning 2000–2007, you'd see a beautiful sight: a gorgeous zig zag sloping upward through that seven-and-a-half-year time frame. It went from under $300 (technically, the bottom was in 2001 at $252), and it went over $670 in early August 2007. But during that time it had some scary, financially bruising periods. That chart would show you no less than seven corrections of 10% to 30%. Ouch! However, that's part of the precious metals world. Bull markets and corrections go together like . . . uh . . . Jekyll and Hyde (sort of).

How would you know the difference? In a word: fundamentals. What are the fundamentals? For precious metals, this is a general reference to things such as supply and demand, economic trends, and other factors that affect the price. Over the span of this decade, investment and industrial demand for gold rose significantly while supply (mined from the earth) did not. As China and India became economic powerhouses, their appetite for gold also grew. The bottom line is that gold's market looks strong for years to come.

A good example of a *bear market* (when an asset or industry has a very long period of dropping prices) is what happened to Japan's real estate market during the 1990s. Real estate values hit an all-time high in early 1991. However, the very expensive prices diminished demand. More houses put on the market increased supply. What followed was 12 consecutive years of falling prices. In late 1991, Japan's housing market didn't faint; it dropped dead.

Watching Political Trends

Gold and silver do have political overtones. The reason being that for thousands of years gold and silver served as money. In recent centuries (especially since the 1930s), governments came to see gold and silver as competing with their own money (currency or fiat money). It became one of those weird relationships you'd see on Dr. Phil. Governments hated gold and silver because they wanted their own currency to dominate, but they also loved gold so their central banks kept a batch of it in safekeeping. The bottom line is that gold can be coveted and confiscated in some political jurisdictions and during certain economic conditions (as was the case in the U.S. during the Great Depression).

How could this affect you in a real way? Say you own stock of a gold-mining company that has significant gold-rich properties in a politically unstable country. What would happen to the value of your stock if that foreign government seized that company's properties?

 Make sure you invest in mining stocks with little or no exposure to hostile or unfriendly political jurisdictions. Generally, governments that are communist, fascist, or socialist are more apt to nationalize (seizure by government force) the property of foreign companies (such as U.S. mining companies). Keep a close eye on those elections! Check out Chapter 19 for more on the politics of precious metals.

Monitoring Inflation

Many precious metals analysts will tell you that the primary catalyst for a precious metals bull market (and natural resources in general) is inflation. History bears this out. Governments may have the power to print more and more money (inflate the money supply otherwise known as *monetary inflation*), but they can't print more gold and silver. When a growing supply of dollars seeks out limited supplies of gold and silver, ultimately gold and silver go up.

As this book is being written, the major countries of the world are increasing (again, "inflating") their money supply at double-digit rates. In an inflationary environment (as we have now), gold and silver (and other precious metals) are sensible parts of a well-balanced portfolio.

Chapter 23

Ten Rules for Metals Traders

*I*f you read nothing else about trading (and you should read everything about it) then at least get the most important points for beginners by checking out this chapter.

Faking Out the Markets First

The cheapest lessons are the ones that you learn without losing a fortune. It almost sounds like "starve a cold and feed a fever" (or was that the opposite?), but it says that the novice or beginning intermediate trader doesn't save up a big chunk of money and jump right into the market while saying "I better get in now before I miss some opportunities." No, there is no hurry. There are trading opportunities every day and in every free market, in good times, bad times, and some mediocre ones as well.

It is better to miss ten good opportunities than to get steamrolled by a bad one. The first thing you should do (besides educating yourself on the market, which I hope that you are doing long before you got to this chapter) is to do some simulated trading.

Also called "paper trading" (for that matter, call it make-believe trading), simulated trading is easier to do than ever before thanks to the Internet and lots of great trading software and services covering the booming financial markets. Apply your ideas, thoughts, theories, and hunches without risking your money. Start with some fake, play money of $10,000 or whatever. Price out some trades and strategies and then track your simulated positions over a few days or a few weeks. How well do you do? More importantly, what did you learn about your choices and how the market behaves?

Having a Plan

You don't have to be General Patton with a grand strategy to proceed. But for crying out loud, have some plan. Ask yourself:

- What am I trying to accomplish?
- What losses am I willing to tolerate?
- What are my strategies to accomplish my trading objectives?

The adage is as old as the hills: Those who fail to plan, plan to fail. Such a phrase was probably coined by that "starve a fever" guy.

A plan is just a disciplined approach in paper (or on your computer) that gives you clear, unemotional guidance about how to go about accomplishing your trading objectives. Trading shouldn't be guesswork (the markets may require some guesswork), but you should have your bases covered so that you know

- When to get out
- When to get in
- When you take profits
- How much losing you will tolerate

It's not that different from when my wife and I would go to a casino. We would play $40 (yeah, big spenders!) at the roulette table. If we lose, we lose $40. If we win a little, then we had some fun. If we are winning nicely, we pull out the original $40 (it can pay for parking and lunch) and maybe another $20 just in case and play the rest for a limited time (for an hour or until show time). It works for us. More importantly, it's a plan we stick to. What's your plan? What are your limits?

Avoiding Committing All Your Cash at Once

I think that this is a big point especially with novice traders. If you have a few thousand and you deploy it immediately, then you can lose it immediately. I spoke to my friend and futures broker, Charlie, and he told me something quite profound. He quoted an observation in the futures industry. The average beginning investor who starts with $5,000 or less in a brokerage account usually loses all of it and closes the account within six months. Yikes!

Yet, if you commit some money now and wait, you could catch more opportunities. If your position is down and you think you chose wisely, then having some extra cash on the sidelines means that you could jump in at a much better price. It's a good idea to stagger the approach so you can take advantage of the market's short-term ups and downs.

Taking Profits Doesn't Hurt

If you are on a hot streak it's a common enticement to "let it ride!" That may not be a bad idea in the world of investing when you can afford to ride out the market's short-term ebbs and flows, but trading means looking for opportunities on a more regular basis. You may end up in a position that will be profitable, and you may agonize about getting out or staying in. That's your call, but the bottom line is that there is no harm in locking in a profit when the market is. That means that you will have cash on the sidelines ready to get back in with the next opportunity.

I have for my students something I call the "100% club." This means that the account is grown to the point that 100% of their original money is taken out and the only money left in the account is 100% profit (we like to call it "house money"). Even if the worse-case scenario hits — losing all the money — does no harm. When times look great, consider taking some money "off the table."

Using Hedging Techniques

Before you leave that message with your gardener, think again. Proficient traders do some things in their trading to protect themselves if the market goes against them. If they buy stock because they expect that stock to go up, they may also buy a protective put just in case the stock goes down. In other words, if you are very bullish about the market or a particular asset, consider hedging; do some things in the account that would do well if you are wrong or if the market temporarily moves against you. A seasoned trader will have on a variety of positions to take advantage of different market outcomes (especially if they are unwelcome outcomes!). Discover all the ways that you can hedge in your account (do your homework and speak to your broker about your various options).

Knowing Which Events Move Markets

Except for sudden, unexpected events (such as a terrorist attack or that visit from Uncle Mo), most market-driving events can be seen coming and you can prepare for them. Events such as government announcements, industry news, political developments, and so on can help you be a more proficient trader.

Financial markets don't act in a vacuum; they frequently react to the world around them. Become an avid news and market observer and keep asking yourself questions that test your logic and common sense. Learn the differences between problems and symptoms and causes and effects. Although markets can seem crazy and irrational in the short term, they ultimately end up acting logically as market conditions unfold. That means that ultimately, your common sense, logic, and market intelligence can help you meet your trading objectives (and make lots of bucks!).

Checking the Trading History

How does your chosen trading vehicle behave, price-wise? Does it have a slow season? What has it done in the past? Also, what has it done in the past under the same conditions today? In trading, you may only be looking forward a few days, weeks, or a few months, but looking backward, you should be reviewing that asset's behavior over years (if possible). Some investment vehicles have trading data going back decades. Because markets tend to be cyclical and they tend to take on characteristics that have happened before, the trading history does matter. Be aware of it and learn to act accordingly.

Using Stop-Loss Strategies

Because the markets do move fast and in unexpected ways, it's good to have a disciplined approach that can "automatically" limit losses. Speak to your broker about what loss-limiting orders and techniques can help enhance your trading proficiency. Many traders will use stop-loss orders or trailing-stop strategies, which can limit the downside. Don't confuse this with hedging which means employing strategies that can be profitable in the event you're wrong or if the market goes against you. Stop-loss strategies are there to get you out of a position before small losses turn into big losses.

Embracing the Experience of Others

Whatever you think of the market or whatever ideas or strategies you may be considering, keep in mind that many people came before you and they have cumulatively seen it all! There are many winners and even more losers and their experiences are there for you to learn from. There are scores of books, Web sites, and newsletters from those who have done it. I include some of my favorites in Chapter 16. If you find out that the majority of successful traders do technical analysis and also employ techniques A, B, and C, then learn from that. What did they do that worked? What mistakes of theirs can you learn from?

Minimizing Transaction Costs

Profitable trading isn't just about what you make; it's also about what you keep. It wasn't that long ago that day-trading was all the rage. But for every day-trader who made a profit, there were hundreds who lost money. The one thing that both winners and losers in the world of day-trading had in common was paying lots of commissions. More trading activity means more commissions and fees. Find ways to cut costs without compromising on your plan and your trading objectives because these costs affect your net profit.

Chapter 24

Ten Ways to Limit Risk

In This Chapter
▶ Getting in . . . slowly
▶ Diversifying in multiple ways
▶ Discovering profitable info

*I*nvesting and speculating isn't just about making money. It's also about limiting losses. Anything you put your money in that trades in a marketplace of buyers and sellers has the chance to go down as well as up. Maximizing the upside while minimizing the risk of loss can be done without much difficulty. Successful investing isn't just what to invest in, it's also the way you do it.

Staggering Your Entry

If you open an account, say stock brokerage account, and your initial investment is $20,000, what will you do? Many investors will probably invest all or most of it immediately. Hey, you sure don't want to miss that gravy train leaving the station for "fabulous-profits-ville"! Do you?

Well, if you're a patient investor and you chose wisely, that's okay. But the reality is that investments zig-zag. Sometimes watching investments is like watching water on a stovetop wondering when it's going to boil. Investments tend to go up a little today and then go down a little tomorrow. Sometimes you get in on a Monday and watch as prices plunge the next day. You then get into the "woulda, coulda, shoulda" of the world of investing. "Gee, had I only waited a day or two I could have gotten in at a much better price!" What's an investor to do?

The odds are overwhelming: You'll probably never buy something at the exact bottom, and you'll never sell something at the exact top. You can only look at that price of that investment on that day and say "Is it worth it for me to buy now?" Much of the market can be unknowable since you can't read the minds of literally millions of investors both big and small. Don't even try. But who said that investing all of your money in one fell swoop immediately was the right way to go?

In the short term, markets today are more volatile and more irrational than at any time in history. Large financial institutions, governments, hedge funds, and other large players are moving large amounts of money — billions and trillions — in and out of the market at lightning speed. With the advent of globalization, the Internet, and other technology, huge amounts of money can zip around markets and economies at the speed of a mouse-click, 24 hours a day. This is why you should stagger your entry.

If, say, you have $10,000 or more to invest, start off investing a quarter or a third of that and wait a few days or a few weeks. Watch your investment and watch the marketplace for news, events, and announcements. If your investment went down and it is still a good asset (say a stock or an ETF), then consider buying more of it. If it was a great investment when you first bought it then it's better as you buy it at a bargain price.

Unless you have a small amount to invest, it's always good to have some investable funds on the sideline — earning interest — waiting for opportunities.

Using Trailing Stops

Although I cover this point in other venues (such as Chapter 16), I think it merits some attention especially in a chapter dealing with risk.

Trailing stops is the active strategy of putting a stop-loss order on your investment (usually a stock or ETF) and adjusting it upward as the investment rises in price. Basically, the purpose of a stop loss is to sell your investment at a specified time if it falls to keep it from losing further. Stop-loss orders don't stop the investment from going up or limit how high it goes; the orders are there to limit the downside.

As I write this in the summer of 2007, there are dozens of mortgage companies that are struggling or going out of business. Many of them are publicly traded and were in many investors' portfolios. Although there were reports around for months that there are massive problems with mortgages (specifically sub-prime mortgages which are mortgages issued to those who have below-average or poor credit), the stock prices of mortgage firms held up well until 2007. Then, you see their stock prices plummet 80% to 95% in literally a few days. This would have been an ideal situation in which trailing could have been used.

Using trailing stops doesn't mean that you were certain that the stocks you held were going to plummet. After all, if you were certain, then you'd definitely sell your stock immediately. Trailing stops are there for those who are not certain about what will happen to the stock in question.

Put yourself in that position for a moment. If you had stock in a mortgage firm and the stock's price was rising yet you are reading about potential problems in the mortgage industry, what would you do? What would a prudent investor do? If that was me, I'd put on a trailing stop so that I can have a measure of protection (and be able to sleep at night).

For precious metals stocks and ETFs, there will come a time when the outlook for them will be uncertain. Not this year or even this decade but sometime in the distant future reports will come out about stagnating or falling demand for precious metals and how supply is increasing because of a mining boom due to sky-high prices. It will be a time when the industry is still feeling ebullient, the same way that the mortgage industry was feeling ebullient in 2005 and early 2006. But when that time comes for your investments, don't get cutesy about it and don't assume that your crystal ball is correct; take some commonsense protective measure such as applying loss-limiting strategies such as the trailing stop.

Diversifying Positions

Being diversified in your investments is certainly addressed elsewhere in this book and in just about every investing how-to book out there. But how could I avoid the topic of diversification in a chapter on limiting risk? It's like trying not to write about fruits and vegetables in a book about avoiding meat.

All that aside, diversifying the positions (such as stocks) in your account is more essential now than it was in the past. Today, the markets are faster and more volatile. Conditions (such as globalization) and challenges (such as terrorism) are more complex and problematic. Large swings up and down by the market indexes are more frequent and more extreme. The moment a problem is reported in an industry you can see a major sell-off occur very rapidly. Having different positions in different stocks will minimize risk.

Diversifying in Markets

Yes, this is a book that unabashedly embraces precious metals. But that doesn't mean to only be in the precious metals market. I mean, I like Elvis Presley too but he is only a small part of my record — uh — CD collection. Allow me to take this small snippet of the book to encourage you to have your money in some other segments of the market that will benefit you and be a good diversification away from just metals.

Of course, in my financial seminars I tell my students to be diversified, but what do you diversify in? A hundred investing experts will tell you a hundred different ideas about what to put your money in. In the realm of stock investing, there are areas to invest in that just make a lot of sense and have proven results over time. I think that every investor, conservative or aggressive, should invest in the stocks of proven companies that produce goods and services that people *need*.

As this book is being written, major events are happening across the global financial markets. The world is overburdened with massive debt and with huge derivative positions. How huge? The world's gross domestic product (GDP) is approximately $45 trillion (as of the end of 2006), yet worldwide derivatives total a staggering $400+ trillion! The United States has a GDP (as of early 2007) of $13 trillion, yet it alone has a debt load of $46 trillion (not including unfunded Social Security and Medicare liabilities of over $50 trillion). These issues don't include other issues that will adversely affect financial markets such as inflation, mortgage-market meltdowns, currency devaluations, terrorism, trade wars, natural disasters, and so on. This complex and uncertain environment poses many risks for the economy in general and the stock market in particular. All of these reasons are why precious metals are high on my list, but, in addition, I want to invest in the *necessities of life*.

In addition to precious metals, consider diversifying into things that the world's burgeoning population needs such as buying stock in proven companies that provide food, water, energy, and other necessities. Since 2000, the general area of natural resources has quietly been outperforming other areas of the investment world, yet the public is barely aware of it. Let me give you an example: water stocks. There are few things more necessary for life than water (outside of oxygen and chocolate, of course). During 2000–2006, investors who put their money in water stocks did much better than investors in tech stocks or Nasdaq or the Dow.

Investing in the necessities of life may not sound sexy or exciting (compared to the latest Internet or biotech stock) but the long-term results can be superior.

Diversify among Different Vehicles

Don't confuse "vehicles" with "markets" (while you're at it, don't confuse investment vehicles with "automobiles"). Stocks, bonds, and options are different investment vehicles. With precious metals, you are best advised to spread your money across different vehicles in the galaxy of precious metals. Look into precious metals bullion, stocks, ETFs, and so on. Read a good book in various precious metals investments. (Hold on . . . you already have one in your hands, so take a look at the table of contents.)

Read the Best Sources

It might cause some suspicion when you hear anyone using the word *best.* To some it is a subjective term: What's best to you may not be best to me. But fortunately, I'm not talking about movies or 17th-century antique furniture. I'm talking about information and advisory services that have consistently offered quality recommendations and research to help you make informed decisions. Different sources have different strengths. For the best source of financial and economic news and data, I like

- Market Watch (www.marketwatch.com)
- Bloomberg (www.bloomberg.com)
- Publications, such as the *Wall Street Journal, Barron's,* and *Investors Business Daily* (and their Web sites)

In the specific realm of precious metals, I want to read sources that specialize in them, have tracked them for a long time, and have a good track record in terms of profitable recommendations. Another big plus is if they generally embrace the economic school of thought that has been the most "on target" with economic analysis and forecasting such as the Austrian school of economics (the best place to learn about this is the Ludwig von Mises Institute at www.Mises.org). Sources that fit this description include analysts, such as:

- Doug Casey
- Jay Taylor
- James Sinclair
- Peter Grandich
- David Morgan
- Howard Ruff
- Harry Schultz

Not including these names in a book about precious metals would be like writing a book about stocks and not mentioning Warren Buffet and Benjamin Graham.

For market news on precious metals, some of the best sources of information, data, and commentary are

- Gold Eagle (www.gold-eagle.com)
- Mine Web (www.mineweb.com)
- The Bullion Desk (www.thebulliondesk.com)
- Kitco's Gold Sheet Links (www.goldsheetlinks.com)

Use your broker as an informational resource. Many brokers can provide e-mail alerts on conditions that you stipulate, such as when a stock hits a certain price level or to provide you alerts on news items such as government announcements, industry developments, and so on.

Using Protective Puts

A put option is a vehicle that you can buy that can go up in value when the asset you are buying it on is dropping in value. Basically, buying a put on some security is like making a bet that it will go down and you get a chance to sell that put at a profit. Buying a put option is speculating, but in this chapter that is not why it is being mentioned. You can use it as a form of insurance called the protective put.

A protective put is an option that you buy on your own stock or ETF. The protective put is a great way to offset the potential downside risk associated with your stock or ETF. Say you own 100 shares of XYZ company stock and its current market value is $50 per share. You are concerned that it may drop in value (for whatever reason) in the near future. You buy a protective put on XYZ so that if the stock goes down in value, the protective put will go up in value. The temporary loss in the stock price could be offset either partially or completely by the profit on the protective put, thereby protecting the value of your position in XYZ stock.

(By the way, I think XYZ stock should be okay. I've seen it mentioned in nearly every stock book I have ever read so I'm sure that lots of people are buying it.) Protective puts are a great way to limit risk. Find out more about puts (protective and otherwise) in Chapter 15.

Avoiding Unstable Political Markets

Finding and tapping into a gold mine is great. But what if it is located in a politically unfriendly jurisdiction? That will be an issue for the health and well-being of your mining stocks. Across the world, more governments view natural resources as strategic assets and are forcibly extending their reach into these areas. In recent years, investors in natural resources (such as precious metals) witnessed these unsettling events. The governments of Russia, Mongolia, and Venezuela (among others) nationalized (the governments seized) properties, such as mines, of foreign companies (many from the U.S.), which resulted in declines of the companies' stock.

It's not difficult to find out if a company has too much exposure to politically unfriendly governments. Generally, the more "free" a country is, the better the political environment. Specifically, you would like a country that has a free market and established private property rights in an English common law framework. Organizations like Freedom House (`www.freedomhouse.org`) keep track of countries and rate them as "free," "partly free," and "not free."

The next things to look at are the company's Web site and annual report, which will tell you where the company's properties are located. After that, it becomes easier to decide what companies are safer than not for your investment dollars.

Looking at Past Trading Patterns

Although I am not usually an aficionado of technical analysis, there are some things about charts that give you some valuable guidance. A stock chart (or one on futures or indexes) that spans the most recent six months, a year, or longer can give you some valuable perspective. You can see the price ranges and patterns that will give you a gauge that measures that investment's volatility and activity. Say you're looking at a mining stock that is at $40 a share and its chart shows that in the past 12 months the stock has swung from $30 a share to $50 in a roller-coaster fashion. You can compare it to the trading patterns of other stocks in general and with stocks in the same industry. This helps you decide if this stock is too volatile (or not volatile enough) for your tastes. It helps you decide where to set stop-loss orders or when it might be time to reduce your exposure.

Taking Some Chips off the Table

It never hurts to take profits. Every now and then, you will have a position that is very profitable. Why not take some profits by cashing in? How would you know when to do it? In some cases, the investment itself will tell you (news and data you come across as you monitor your investment). In other cases, your own personal situation may tell you. Do you have or can you see coming some financial needs? If you have a profitable stock and you noticed that your credit card bill has gotten too high, perhaps that is a good signal to cash out some of your winning positions.

Index

BUSINESS, CAREERS & PERSONAL FINANCE

0-7645-9847-3   0-7645-2431-3

Also available:
- Business Plans Kit For Dummies
  0-7645-9794-9
- Economics For Dummies
  0-7645-5726-2
- Grant Writing For Dummies
  0-7645-8416-2
- Home Buying For Dummies
  0-7645-5331-3
- Managing For Dummies
  0-7645-1771-6
- Marketing For Dummies
  0-7645-5600-2

- Personal Finance For Dummies
  0-7645-2590-5*
- Resumes For Dummies
  0-7645-5471-9
- Selling For Dummies
  0-7645-5363-1
- Six Sigma For Dummies
  0-7645-6798-5
- Small Business Kit For Dummies
  0-7645-5984-2
- Starting an eBay Business For Dummies
  0-7645-6924-4
- Your Dream Career For Dummies
  0-7645-9795-7

HOME & BUSINESS COMPUTER BASICS

0-470-05432-8   0-471-75421-8

Also available:
- Cleaning Windows Vista For Dummies
  0-471-78293-9
- Excel 2007 For Dummies
  0-470-03737-7
- Mac OS X Tiger For Dummies
  0-7645-7675-5
- MacBook For Dummies
  0-470-04859-X
- Macs For Dummies
  0-470-04849-2
- Office 2007 For Dummies
  0-470-00923-3

- Outlook 2007 For Dummies
  0-470-03830-6
- PCs For Dummies
  0-7645-8958-X
- Salesforce.com For Dummies
  0-470-04893-X
- Upgrading & Fixing Laptops For Dummies
  0-7645-8959-8
- Word 2007 For Dummies
  0-470-03658-3
- Quicken 2007 For Dummies
  0-470-04600-7

FOOD, HOME, GARDEN, HOBBIES, MUSIC & PETS

0-7645-8404-9   0-7645-9904-6

Also available:
- Candy Making For Dummies
  0-7645-9734-5
- Card Games For Dummies
  0-7645-9910-0
- Crocheting For Dummies
  0-7645-4151-X
- Dog Training For Dummies
  0-7645-8418-9
- Healthy Carb Cookbook For Dummies
  0-7645-8476-6
- Home Maintenance For Dummies
  0-7645-5215-5

- Horses For Dummies
  0-7645-9797-3
- Jewelry Making & Beading For Dummies
  0-7645-2571-9
- Orchids For Dummies
  0-7645-6759-4
- Puppies For Dummies
  0-7645-5255-4
- Rock Guitar For Dummies
  0-7645-5356-9
- Sewing For Dummies
  0-7645-6847-7
- Singing For Dummies
  0-7645-2475-5

INTERNET & DIGITAL MEDIA

0-470-04529-9   0-470-04894-8

Also available:
- Blogging For Dummies
  0-471-77084-1
- Digital Photography For Dummies
  0-7645-9802-3
- Digital Photography All-in-One Desk Reference For Dummies
  0-470-03743-1
- Digital SLR Cameras and Photography For Dummies
  0-7645-9803-1
- eBay Business All-in-One Desk Reference For Dummies
  0-7645-8438-3
- HDTV For Dummies
  0-470-09673-X

- Home Entertainment PCs For Dummies
  0-470-05523-5
- MySpace For Dummies
  0-470-09529-6
- Search Engine Optimization For Dummies
  0-471-97998-8
- Skype For Dummies
  0-470-04891-3
- The Internet For Dummies
  0-7645-8996-2
- Wiring Your Digital Home For Dummies
  0-471-91830-X

* Separate Canadian edition also available
† Separate U.K. edition also available

Available wherever books are sold. For more information or to order direct: U.S. customers visit www.dummies.com or call 1-877-762-2974. U.K. customers visit www.wileyeurope.com or call 0800 243407. Canadian customers visit www.wiley.ca or call 1-800-567-4797.

SPORTS, FITNESS, PARENTING, RELIGION & SPIRITUALITY

0-471-76871-5   0-7645-7841-3

Also available:
- Catholicism For Dummies
  0-7645-5391-7
- Exercise Balls For Dummies
  0-7645-5623-1
- Fitness For Dummies
  0-7645-7851-0
- Football For Dummies
  0-7645-3936-1
- Judaism For Dummies
  0-7645-5299-6
- Potty Training For Dummies
  0-7645-5417-4
- Buddhism For Dummies
  0-7645-5359-3

- Pregnancy For Dummies
  0-7645-4483-7 †
- Ten Minute Tone-Ups For Dummies
  0-7645-7207-5
- NASCAR For Dummies
  0-7645-7681-X
- Religion For Dummies
  0-7645-5264-3
- Soccer For Dummies
  0-7645-5229-5
- Women in the Bible For Dummies
  0-7645-8475-8

TRAVEL

0-7645-7749-2   0-7645-6945-7

Also available:
- Alaska For Dummies
  0-7645-7746-8
- Cruise Vacations For Dummies
  0-7645-6941-4
- England For Dummies
  0-7645-4276-1
- Europe For Dummies
  0-7645-7529-5
- Germany For Dummies
  0-7645-7823-5
- Hawaii For Dummies
  0-7645-7402-7

- Italy For Dummies
  0-7645-7386-1
- Las Vegas For Dummies
  0-7645-7382-9
- London For Dummies
  0-7645-4277-X
- Paris For Dummies
  0-7645-7630-5
- RV Vacations For Dummies
  0-7645-4442-X
- Walt Disney World & Orlando
  For Dummies
  0-7645-9660-8

GRAPHICS, DESIGN & WEB DEVELOPMENT

0-7645-8815-X   0-7645-9571-7

Also available:
- 3D Game Animation For Dummies
  0-7645-8789-7
- AutoCAD 2006 For Dummies
  0-7645-8925-3
- Building a Web Site For Dummies
  0-7645-7144-3
- Creating Web Pages For Dummies
  0-470-08030-2
- Creating Web Pages All-in-One Desk
  Reference For Dummies
  0-7645-4345-8
- Dreamweaver 8 For Dummies
  0-7645-9649-7

- InDesign CS2 For Dummies
  0-7645-9572-5
- Macromedia Flash 8 For Dummies
  0-7645-9691-8
- Photoshop CS2 and Digital
  Photography For Dummies
  0-7645-9580-6
- Photoshop Elements 4 For Dummies
  0-471-77483-9
- Syndicating Web Sites with RSS Feeds
  For Dummies
  0-7645-8848-6
- Yahoo! SiteBuilder For Dummies
  0-7645-9800-7

NETWORKING, SECURITY, PROGRAMMING & DATABASES

0-7645-7728-X   0-471-74940-0

Also available:
- Access 2007 For Dummies
  0-470-04612-0
- ASP.NET 2 For Dummies
  0-7645-7907-X
- C# 2005 For Dummies
  0-7645-9704-3
- Hacking For Dummies
  0-470-05235-X
- Hacking Wireless Networks
  For Dummies
  0-7645-9730-2
- Java For Dummies
  0-470-08716-1

- Microsoft SQL Server 2005 For Dummies
  0-7645-7755-7
- Networking All-in-One Desk Reference
  For Dummies
  0-7645-9939-9
- Preventing Identity Theft For Dummies
  0-7645-7336-5
- Telecom For Dummies
  0-471-77085-X
- Visual Studio 2005 All-in-One Desk
  Reference For Dummies
  0-7645-9775-2
- XML For Dummies
  0-7645-8845-1

HEALTH & SELF-HELP

0-7645-8450-2

0-7645-4149-8

Also available:
- Bipolar Disorder For Dummies
0-7645-8451-0
- Chemotherapy and Radiation
For Dummies
0-7645-7832-4
- Controlling Cholesterol For Dummies
0-7645-5440-9
- Diabetes For Dummies
0-7645-6820-5* †
- Divorce For Dummies
0-7645-8417-0 †

- Fibromyalgia For Dummies
0-7645-5441-7
- Low-Calorie Dieting For Dummies
0-7645-9905-4
- Meditation For Dummies
0-471-77774-9
- Osteoporosis For Dummies
0-7645-7621-6
- Overcoming Anxiety For Dummies
0-7645-5447-6
- Reiki For Dummies
0-7645-9907-0
- Stress Management For Dummies
0-7645-5144-2

EDUCATION, HISTORY, REFERENCE & TEST PREPARATION

0-7645-8381-6

0-7645-9554-7

Also available:
- The ACT For Dummies
0-7645-9652-7
- Algebra For Dummies
0-7645-5325-9
- Algebra Workbook For Dummies
0-7645-8467-7
- Astronomy For Dummies
0-7645-8465-0
- Calculus For Dummies
0-7645-2498-4
- Chemistry For Dummies
0-7645-5430-1
- Forensics For Dummies
0-7645-5580-4

- Freemasons For Dummies
0-7645-9796-5
- French For Dummies
0-7645-5193-0
- Geometry For Dummies
0-7645-5324-0
- Organic Chemistry I For Dummies
0-7645-6902-3
- The SAT I For Dummies
0-7645-7193-1
- Spanish For Dummies
0-7645-5194-9
- Statistics For Dummies
0-7645-5423-9

Get smart @ dummies.com®

- **Find a full list of Dummies titles**
- **Look into loads of FREE on-site articles**
- **Sign up for FREE eTips e-mailed to you weekly**
- **See what other products carry the Dummies name**
- **Shop directly from the Dummies bookstore**
- **Enter to win new prizes every month!**

* Separate Canadian edition also available
† Separate U.K. edition also available

Available wherever books are sold. For more information or to order direct: U.S. customers visit www.dummies.com or call 1-877-762-2974.
U.K. customers visit www.wileyeurope.com or call 0800 243407. Canadian customers visit www.wiley.ca or call 1-800-567-4797.

Do More with Dummies

Tickle my ribs!

Grilling FOR DUMMIES

12" x 12" kit
Scrapbooking Basics FOR DUMMIES
Create unforgettable scrapbooking pages faster and easier than ever
Scrapbooking — the Fun & Easy Way

Sewing Patterns FOR DUMMIES
Quilting Notions FOR DUMMIES

Cocktail Kit FOR DUMMIES
Everything you need to shake, measure, and serve like the pros
Get dozens of hip and classy recipes

Poker FOR DUMMIES

Golf FOR DUMMIES

Pilates Workout FOR DUMMIES

'80s Pop Music FOR DUMMIES

'70s Soul Music FOR DUMMIES

Wall & Ceiling Repair Kit FOR DUMMIES

Tarot Deck & Book Set FOR DUMMIES
Tarot for the Rest of Us!

Sudoku FOR DUMMIES The Game
Includes more than 50 Sudoku puzzles for hours of fun

Texas Hold 'em FOR DUMMIES
Texas Hold 'em FOR DUMMIES
A Card Game for the Rest of Us!

**Instructional DVDs • Music Compilations
Games & Novelties • Culinary Kits
Crafts & Sewing Patterns
Home Improvement/DIY Kits • and more!**

Check out the Dummies Specialty Shop at www.dummies.com for more information!

WILEY